Developing a Turnaround Business Plan

Norton Paley

Developing a Turnaround Business Plan

Leadership Techniques to Activate Change Strategies, Secure Competitive Advantage, and Preserve Success

CRC Press is an imprint of the
Taylor & Francis Group, an **informa** business

A PRODUCTIVITY PRESS BOOK

CRC Press
Taylor & Francis Group
6000 Broken Sound Parkway NW, Suite 300
Boca Raton, FL 33487-2742

© 2016 by Norton Paley
CRC Press is an imprint of Taylor & Francis Group, an Informa business

No claim to original U.S. Government works

Printed on acid-free paper
Version Date: 20150213

International Standard Book Number-13: 978-1-4987-0590-5 (Hardback)

This book contains information obtained from authentic and highly regarded sources. Reasonable efforts have been made to publish reliable data and information, but the author and publisher cannot assume responsibility for the validity of all materials or the consequences of their use. The authors and publishers have attempted to trace the copyright holders of all material reproduced in this publication and apologize to copyright holders if permission to publish in this form has not been obtained. If any copyright material has not been acknowledged please write and let us know so we may rectify in any future reprint.

Except as permitted under U.S. Copyright Law, no part of this book may be reprinted, reproduced, transmitted, or utilized in any form by any electronic, mechanical, or other means, now known or hereafter invented, including photocopying, microfilming, and recording, or in any information storage or retrieval system, without written permission from the publishers.

For permission to photocopy or use material electronically from this work, please access www.copyright.com (http://www.copyright.com/) or contact the Copyright Clearance Center, Inc. (CCC), 222 Rosewood Drive, Danvers, MA 01923, 978-750-8400. CCC is a not-for-profit organization that provides licenses and registration for a variety of users. For organizations that have been granted a photocopy license by the CCC, a separate system of payment has been arranged.

Trademark Notice: Product or corporate names may be trademarks or registered trademarks, and are used only for identification and explanation without intent to infringe.

Library of Congress Cataloging-in-Publication Data
Paley, Norton.
Developing a turnaround business plan : leadership techniques to activate change strategies, secure competitive advantage, and preserve success / Norton Paley.
pages cm
Includes index.
ISBN 978-1-4987-0590-5 (alk. paper)
1. Corporate turnarounds. 2. Organizational change. 3. Business planning. I. Title.
HD58.8.P345 2016
658.4'012--dc23   2015002235

Visit the Taylor & Francis Web site at
http://www.taylorandfrancis.com

and the CRC Press Web site at
http://www.crcpress.com

To the memory of Rebbe Nachman

Contents

Introduction xiii

SECTION I   Developing a Turnaround Business Plan

Chapter 1   Identify the Root Causes That Trigger a Turnaround 3

    The Primary Conditions That Activate a Turnaround 3
    Types of Competitive Campaigns 7
    Physical and Psychological Characteristics of a
    Competitive Conflict 16
        Conflicts Do Not Break Out Unexpectedly 16
        Conflicts Require Neutralizing the Competitor 17
        Conflicts Are Not Isolated Events 19
        Campaigns Cannot Be Interrupted 20
    Factors That Can Bring a Campaign to a Standstill 22
        Competitive Conflicts Contain Elements of Chance 25

Chapter 2   Prepare the Organization for a Turnaround 27

    Introduction 27
    The Physical Dimension 28
    The Psychological Dimension 29
        Organizational Culture 29
        Seek Maximum Input from All Levels of Employees 32
            Stay on the Offensive 33
            Act as an Aggressive Competitor 35
            Build a Strong Market Position 35
            Stay Close to Evolving Technology 35
            Establish Strong Internal Communications 36
    Strong versus Weak Cultures 37
    The Power of Morale 37
    Relationships between Leader and Staff: Expectations
    for Developing a Turnaround Plan 44
        Expect Active Participation from Staff 44

Expect Staff to Maintain Momentum 45
Expect Staff to Neutralize Competitor's Strategies 47
Expect Innovative Thinking 49
Expect Staff to Stay Alert to Competitive and Market
Conditions 50
Expect Staff to Respond to Negative Behavior 51
References 52

Chapter 3 Prepare a Turnaround Strategy Plan 53

Introduction 53
Establishing a Strategic Direction 55
Objectives 60
Strategies 62
Postcampaign Strategies to Secure a Turnaround 65
    Signs of Complacency 66
    Signs of Inflexibility During a Time of Disruptive
    Change 66
    Signs of Lethargy 67
    Signs of Unnecessary Dispersal of Resources 67
    Signs of Inadequate Competitor Intelligence 68
    Signs of an Anemic Corporate Culture 68
    Signs of Ineffectual Leadership 69
    Signs of Sagging Morale 69
    Signs of Failure to Apply the Principles of Strategy 70
        Primary Strategies 70
        Supporting Strategies 72

SECTION II Activate Change Strategies

Chapter 4 Leadership Techniques to Activate a Turnaround 79

Introduction 79
The Transforming Effect of Courage 81
    Activating Intuition 83
    The Power of Determination 85
    Presence of Mind 87
    Honor, Recognition, and Reputation 87
    Strength of Mind 89

Strength of Character 90
　　Leadership Applied to Market Selection 92
　　　Natural Markets 93
　　　Leading Edge Markets...93
　　　Key Markets 94
　　　Linked Markets 94
　　　Central Markets 95
　　　Challenging Markets 95
　　　Difficult Markets 96
　　　Encircled Markets 96
　　Intellectual Standards and Performance 97

Chapter 5 The Competitive Campaign: Structure and
　　　Characteristics 99

　　Introduction 99
　　The Essential Components of a Campaign 104
　　　Duration of a Campaign 106
　　　Conducting the Campaign 107
　　　Defense versus Offense..109
　　　The Characteristics of Offense 111
　　　Campaign Follow-Up 112
　　　Use of Reserves...114

SECTION III　The Essential Elements of Turnaround Strategies

Chapter 6 Bold Action versus Cautious Restraint......................... 119

　　Introduction 119
　　Applying Boldness.. 123
　　Finding Decisive Points 128
　　Applying Caution 128
　　Management Tools for Decision Making.............................. 131
　　　BCG Growth-Share Matrix 132
　　　General Electric Business Screen.. 134
　　　Arthur D. Little Matrix..136
　　　Management by Objectives (MBO).......................................138
　　　Six Sigma 139

Chapter 7 Concentration versus Dispersal Strategy 141

Introduction 141
Implementing a Concentration Strategy 144
   Consumers 144
   Intermediaries 146
      Introducing a New Product 146
      Intensifying Market Coverage 147
      Making a Smooth Transition When Adding or
      Replacing Distributors 147
      Changing Methods of Distribution to
      Complement Changes in Business Strategy and
      Movements in the Industry 149
   Competitors 150
   Regulatory Issues and Industry Trends 151
   Leadership and Management 152
      Market Research 152
      Planning 153
      Organization 154
Guidelines to Utilizing a Concentration Strategy 155
Dispersal Strategy 156
Utilizing Agents for Competitive Intelligence 159
   General Agents 160
   Inside Agents 160
   Double Agents 161
   Expendable Agents 162
   Living Agents 162
How to Conduct a SWOT Analysis 163

Chapter 8 Indirect versus Direct Strategy 167

Introduction 167
   Think Strategically 169
   Maneuver Tactically 173
   Unbalance the Competitor 181
      Stress 182
      Fear 183

Chapter 9 Valuing Surprise and Speed 187
    Introduction 187
    Speed..191
    Positioning..194
        Develop a Positioning Strategy 195
        Branding..197
    Barriers to Implementing Speed..198
        Leadership...198
        The Organization 199
        The Ending Point 200
    Valuing Surprise and Speed 204

SECTION IV Secure Competitive Advantage and Preserve Success

Chapter 10 Use a Pretest to Evaluate Your Turnaround Plan for Competitive Advantage 209
    Introduction 209
        Conditions Triggering a Turnaround 210
        The Organization 211
        The Turnaround Strategy Plan.............................212
        Leadership...214
        The Competitive Campaign 215
        Bold Action 217
        Concentration versus Dispersal Strategy 219
        Indirect versus Direct Strategy 220
        Surprise and Speed 221
    A Final Word.. 223

Index.. 227
About the Author... 243

Introduction

A number of prominent companies made business headlines during a recent 12-month period. All shared one common theme: They were involved in various stages of developing and implementing a turnaround plan.

Kodak, General Motors, and BlackBerry were working on turning around their respective businesses after coming out of bankruptcy. Intel, Levi Strauss, Yahoo, J.C. Penney, Hewlett-Packard, Coca Cola, and Staples were experiencing declines in revenues and threats to their market share positions. They, too, were immersed in developing turnaround plans.

At Samsung, the CEO sent an email to all employees: *You must do better. We must start anew to reach loftier goals and ideals.* That terse message was sent after the company recorded a banner year in 2013. He recognized that the good times could not last; competitors could be readying plans to attack the company at any time. The intent of the directive was to maintain an ongoing state of urgency.

Even the CEO's tone was firm enough to serve as marching orders to take action. The meaning being: prepare competitive plans, activate fresh strategies, secure competitive advantage, and do everything to preserve hard-won success. In its broadest interpretation, such orders meant that every manager at the division, business unit, and product-line levels had to work on developing and maintaining a turnaround against all possible contingencies.

In another instance, a group of senior executives of lesser known organizations revealed their problems in a survey conducted by a respected business publication. Here's a sampling:

"We're facing an avalanche of competition and are bogged down."
"My company is losing momentum and losing its way."
"Need to find the way back to the market where we got creamed."
"Need to develop an organization to yield a new and far more competitive company."
"Sluggish markets, ferocious competition, and our management's own ineptness pummeled the organization."

"We need to keep the company from becoming an also-ran in the industry."

"An element of arrogance in our company's culture leads some managers to ignore the intense competition."

Do these problems sound like quick-fix issues? Are those organizations at the stage of calling in turnaround specialists and consulting firms? More likely a comprehensive internal process is needed initially to (1) stop the predicament from deteriorating further, (2) identify the root causes that triggered the problem, (3) gain the active participation of personnel to develop a turnaround business plan, and (4) implement the plan with the aims of securing a competitive advantage and preserving success.

This book, then, is directed to managers who recognize the need to set in motion a turnaround plan before a crisis hits their individual areas of responsibility. Whereas the primary output is a turnaround strategy, the focus is on honing skills to think with greater precision about managing people, money, and physical resources in a competitive conflict.

To that end, turnaround goes beyond the obligatory ritual of initiating austerity and scrutinizing every cost item. Such defensive moves are but pauses to regroup and prepare for offensive action. Thus, the following chapters delve into the mindsets and the thought processes to sort out the myriad psychological issues that go into the why, and how personnel react to conflict. They address the corporate culture and business unit subcultures that form the underpinnings for all subsequent actions.

As a handbook and reference guide, it serves those managers who are charged with resolving a competitive crisis and restoring the equilibrium of an organization or business unit. In part, that means determining if it will stay in the same market and be able to defend itself, or move in a new direction and change its business model, based on the strategic direction of the plan.

Beyond the structuring of a viable turnaround business plan, a major portion of the book is devoted to strategy. This includes sections about neutralizing the competitor's capabilities through strategy and maneuver, making a campaign more costly for a competitor, and causing the opposing manager to spread his resources and become vulnerable. Other topics discuss how to

- develop a flexible organizational capability for rapid response to competitive countermoves;

- identify the characteristics of offensive and defensive strategies, and the advantages of each;
- use competitive intelligence to identify decisive points;
- determine the type of campaign and decide the ending point of the effort; and
- assess the impact of leadership styles on implementing competitive strategies.

ORGANIZATION OF THE BOOK

Section I: Developing a Turnaround Business Plan

Chapter 1: Identify the Root Causes That Trigger a Turnaround

The book opens with an examination of the primary causes underlying a competitive crisis. It distinguishes among the types of competitive campaigns, and differentiates the physical and psychological characteristics of a competitive conflict. A review of the factors shows how a campaign can come to a standstill with faulty intelligence, fear, inept leadership, inadequate effort, inflexibility, friction, and complacency. In turn, those factors introduce the realities of competitive conflicts when coping with the inevitable presence of chance and luck.

Chapter 2: Prepare the Organization for a Turnaround

This chapter discusses the necessity of implementing a top-to-bottom planning system to integrate the thinking and active participation of personnel with diverse experiences and backgrounds. It provides special focus on a healthy corporate culture and heightened morale as essential factors in preparing the organization for a turnaround.

Chapter 3: Prepare a Turnaround Strategy Plan

This chapter goes into the organization and design of a well thought-out turnaround plan. It details the development of a strategic direction as the focal point for the objectives and turnaround strategies that follow. Also included is the process for shaping postcampaign strategies to secure a turnaround.

Section II: Activate Change Strategies

Chapter 4: Leadership Techniques to Activate a Turnaround

The chapter lists the strategic issues on which a superior leader must focus, including bolstering the organization with the motivation and cohesive spirit to reach its goals. It identifies the key qualities essential to the mind of the leader: intuition, ability to manage doubt, stress, uncertainty, and chance. It deals with the attributes of intelligence, courage, determination, creativity, and the faculty to see the entire strategic picture.

Chapter 5: The Competitive Campaign: Structure and Characteristics

Included are the causes of friction that can bring a campaign to a grinding halt, and the actions needed to reduce their damaging effects. Also covered are the essential components of a campaign and the characteristics that distinguish the defensive campaign from an offensive one.

Section III: The Essential Elements of Turnaround Strategies

Chapter 6: Bold Action versus Cautious Restraint

Using case examples, this chapter discusses the various attributes and applications of bold action in a competitive campaign compared with a more cautious approach. To assist in making correct choices, the classic management tools for decision making are included. Also listed are the types of competitive intelligence that are needed when planning bold moves against rivals.

Chapter 7: Concentration versus Dispersal Strategy

Where the tendency is to avoid risk by spreading resources over numerous areas, the advantages of concentration are provided. Also, guidelines provide techniques for utilizing a concentration strategy.

Chapter 8: Indirect versus Direct Strategy

This chapter shows how to incorporate indirect strategies into a turnaround plan and, thereby, be able to maneuver around competitor's strong points. It also discusses the physical and psychological components of an indirect strategy.

Chapter 9: Valuing Surprise and Speed

The two primary purposes of initiating speed and surprise in a competitive campaign are to first achieve superiority at a decisive point, and, secondly, neutralize the rival's capabilities that would prevent achieving the planned objectives. The chapter includes the techniques for using speed and surprise, as well as the barriers that prevent implementing them. The chapter also includes approaches to determining the ending point of a campaign.

Section IV: Secure Competitive Advantage and Preserve Success

Chapter 10: Use a Pretest to Evaluate Your Turnaround Plan for Competitive Advantage

A set of nine guidelines pinpoint the key factors for pretesting the effectiveness of the turnaround plan. With such topics as corporate culture, morale, leadership, competitive intelligence, organizational design, and campaign strategy, the pretest also provides a review of those factors that would preserve success and become part of a poststrategy.

Section I

Developing a Turnaround Business Plan

- Identify the Root Causes That Trigger a Turnaround
- Prepare the Organization for a Turnaround
- Prepare a Turnaround Strategy Plan

1

Identify the Root Causes That Trigger a Turnaround

Chapter Objectives

Be able to

1. list the market conditions that activate a turnaround plan;
2. distinguish among the various types of competitive campaigns;
3. differentiate among the physical and psychological characteristics that affect morale, create friction within the rank-and-file, and form positive or negative behavioral reactions;
4. identify the factors that can bring a campaign to a standstill; and
5. recognize how chance and luck are natural components of competitive conflicts and turnarounds.

THE PRIMARY CONDITIONS THAT ACTIVATE A TURNAROUND

The most visible causes of a competitive crisis occur when a low-cost competitor enters a market through aggressive pricing, or where a rival organization attracts marketplace attention by employing new technologies. The following two company examples illustrate these common causes.

Panasonic, the Japanese electronics giant, had employed a market strategy of trying to be all things to all customers, making everything from smartphones to solar panels. In the two years ending 2013, the Osaka-based company lost $148 million, which the company attributed to bold competition from low-cost manufacturers in South Korea and China.

In response, Panasonic followed a familiar pattern that other organizations with similar problems had taken to solve the crisis. It began its turnaround by chopping away at costs and exiting money-losing businesses. The company moved out of panels for plasma TVs, trimmed circuit board manufacturing, and gave up on developing consumer smartphones. Overall, Panasonic sharply reduced reliance on consumer electronics, where it had trailed behind Samsung Electronics and Apple.

Eastman Kodak faced a different crisis. Thinking the company was protected by a proud name, a distinguished market history, and a strong public image, its overly confident executives lagged behind in switching to a new industry technology: digital photography.

Any substantive action Kodak attempted was too little, too late. It subsequently declared bankruptcy in 2013, during which time it reorganized and downsized into a fraction of its original size, exited its primary markets, and moved in an entirely new direction.

The irony is that Kodak saw the coming of digital photography. It is even credited with inventing the digital camera in 1975. In the end, however, Kodak management remained focused on protecting its old technology and stoically watched as its market presence declined. The result? An industry that is identified with Kodak was relinquished to aggressive and forward-looking rivals.

The problems that characterized these two organizations are symptomatic of the competitive crisis facing many organizations that get stuck in untenable situations. Some develop turnarounds and are successful in pulling away from their stalled positions. Others fall away and disappear from the market altogether.

One such company that attempted to reverse a potential competitive crisis is Intel Corporation. The giant chip maker proudly claimed that its processors ran 8 out of 10 PCs in the world. The problem is that the demand for PCs is diminishing. PC unit shipments dropped 4-percent in 2012, followed by a 10-percent drop in 2013, with a forecast of a continued decline. The reason is that the world has shifted to mobile with the massive acceptance of smartphones and tablets that appeal to virtually every user segment.

The further problem was that, as of 2013, Intel represented just 1 percent market share in tablets and phones. According to analysts, the company also was caught off guard by the speed of the transition and the ensuing demand for new designs in low-power chips for mobile devices.

The company's central focus had been on its core high-performance chips designed for machines that plug into a wall.

Once the Intel Board internalized the full magnitude of the impending crisis, it moved quickly to initiate a rapid turnaround plan. A new CEO and president were selected. What immediately followed was a complete management shakeup. The new executive team quickly invited other senior-level executives into a strategy meeting where each was given an urgent assignment to submit an action plan pointed to a single directive: Move swiftly and decisively into smartphones and tablets.

In still other organizations, obvious moves dominate executives' ideas for a turnaround: cut costs, eliminate product lines, shelve new product development, exit markets, and attempt change through other knee-jerk activities. However, they are but partial moves and do not assure a successful turnaround plan.

Lurking below the surface are not-so-obvious questions that should be answered before any meaningful turnaround plan can be formed. And, within these questions are clues to a set of activities that need to be addressed:

> Was the approaching competitive threat wholly unexpected, or could it have been avoided if there were a streamlined organization that permitted the flow of intelligence from the field to the right decision-making managers who could have responded rapidly to the danger?
>
> Could a properly developed strategic business plan have averted the crisis and turned around the situation, assuming it contained a long-term strategic direction based on substantive information that included contingency scenarios and not subject to the emotional reactions of the moment?
>
> To what extent had the staff internalized the seriousness of facing a combative conflict, which require astute strategies and maneuvers to neutralize the rival's capabilities?
>
> Were key members of the staff—including sales staff*—trained in developing competent strategies for going on the offensive to expand a market presence and to defend what has been achieved?

---

* The assumption is that the sales staff is, or should be, viewed as general managers of their respective sales territories. Consequently, they should be responsible and accountable for recognizing an impending competitive crisis and be able to recommend responsive strategies. Several insightful organizations have taken the lead in orienting and training their sales staff to elevate their thinking to a strategic level.

Did cross-functional teams exist at various levels with responsibilities to develop business plans, and did team members understand that their collective actions must complement the corporate culture for plans to succeed?

Did management recognize and take into account how unexpected forms of market resistance, friction, and the vagaries of chance could bring a turnaround plan to a standstill, and did it grasp the need for flexibility as an essential component of management practice?

Did the senior executives and line managers display the leadership qualities that would inspire individuals in conflict, did they maintain the strength of mind at times of stress, and did they display the determination and boldness to move forward and implement a plan with a winning attitude?

Did managers understand that a competitive crisis is not solved in a single campaign, rather, it is made up of a series of linked events, each leading to the next, so that one campaign should not be viewed with any sense of finality?

To what extent did the organization and its personnel maintain unity even in crisis, so that it could support a turnaround with a competitive spirit and high morale?

Did the managers fully comprehend an essential aim of strategy: uncover those decisive junctures that represent the focal point, which would then become the primary objectives of the turnaround plan?

Did the leaders understand the importance of identifying the ending point of the effort, that is, the point at which further expenditures of resources would be counterproductive?

Was there a poststrategy in place to secure success, that is, to set in motion contingency plans to actively defend a market position against a competitive attack?

What are all these questions about?

First and foremost they emphatically imply that a competitive crisis is not some random occurrence that suddenly springs upon a company and surprises a manager—or it certainly shouldn't do so.

Second, there assumes underlying, endemic flaws that have been festering for long periods of time embedded within the organizational or cultural makeup of the firm, so that a sudden aggressive move by a competitor could explode and disable an organization's ability to beat off a competitive attack.

Third, where there is a strong likelihood of a competitive crisis, the object of a turnaround plan is to lessen the possibilities of failure and, to a greater extent, increase the potential for success.

Fourth, concurrent with the previous point, the further object of the turnaround plan is to neutralize the competitor's advantage and even go as far as to defuse his capabilities to do any further harm.

Fifth, numerous types of competitive campaigns exist; each has a distinctive purpose and requires a customized action plan with objectives aligned to the strategic business plan and the organization's culture.

These issues and others represent the content of the chapters that follow.

TYPES OF COMPETITIVE CAMPAIGNS

As to the fifth point above, the various types of competitive campaigns are reported in the business press in a variety of forms, such as:

- Walmart attacks one of the largest consumer-electronics chains, Best Buy.
- BMW shifts production to lightweight carbon fiber ahead of competitors to give its electric cars a performance advantage.
- Google launches a new computer operating system and attacks Microsoft.
- BASF challenges Monsanto's dominance in the global seed market.

Although these headline descriptions may not have reached a crisis stage for those organizations under attack, they nevertheless do call for the defending firms to develop action plans that include turnaround scenarios.

Each of the above competitive situations, therefore, take on its own unique character. On the surface, the actions appear as straightforward competitive confrontations to gain an advantageous position, disrupt a competitor, or reverse declining sales by using a variety of thrusts, such as introducing cutting-edge technology, offering enhanced services, launching new products, initiating low prices, and the like.

Yet there are numerous other types of competitive campaigns, such as the following.

Campaigns to Reclaim a Former Market Position—In this type of campaign, an organization could have a difficult time coming up with a workable turnaround strategy. One major hurdle is motivating personnel. It is the leadership's absolute priority to regain their confidence and bolster morale. And, if the company aims to successfully reclaim its former position, yet waits too long before taking action, the defending competitor can consolidate its position into a formidable barrier. On the other hand, there is an advantage for the attacking company that has once been in possession of the market to know from previous experience where the decisive points on which to focus attention.

A classic case is Xerox, which still serves as a constructive lesson for today's challenge to reclaim a former market position. During the 1970s, the company concentrated almost entirely on selling its large copiers to big companies.

Whether by choice or by management's myopic view of the total marketplace, the company was blindsided as it attempted to protect its dominant market position. In so doing, Xerox managers left a huge gap for enterprising Japanese copier makers to snatch an advantage by entering an unattended market with their small copiers to reach the huge numbers of small and midsize companies.

Once established, aggressive rivals, such as Canon and Ricoh, took the natural route by expanding their lines of copiers, which finally encroached on Xerox's hold of the large company segment. The subsequent precipitous drop in Xerox's market share has taken decades to recover through numerous campaigns. Meanwhile the intruders embedded themselves solidly in the North American market.

Defensive Campaigns to Retain a Share of Market in a Key Region—In this type of campaign, the defender initially has the advantage. However, it cannot be a passive defense. There must be an action component to the strategy. The defender can observe the type of strategy that the attacking competitor uses and then be in a better position to take counteraction to disrupt the rival's plans.

This type of campaign is somewhat similar to the situation faced by Best Buy. The retailer had been steadily losing its market position and was even shunned by market analysts who were ready to write off the company. During 2013, however, the company began a remarkable turnaround, if only measured by one metric, its stock, which gained nearly 240 percent, placing Best Buy among the top three performing stocks in the Standard & Poor's 500-stock index.

The change began with newly appointed CEO Hubert Joly. At the time he took over, Best Buy's staff was in turmoil, demoralized from declining same-store sales; its stock price in free fall; and the previous chief executive ousted in a sex scandal. The French-born executive analyzed Best Buy's strengths and weaknesses and found that many shoppers were browsing in his stores and then using their smartphones to compare prices and order online from competitors.

Joly acknowledged that price was the central problem and moved to match the lowest price by cutting costs and streamlining operations. He correctly noted, too, that price was not the deciding factor in a sale. He thereby formulated a strategy that centered on superior advice, convenience, and service to spearhead his turnaround against competitors. He wanted to make certain that a customer who came to Best Buy's store had no reason to buy elsewhere.

Thus, with the primary objective of converting store traffic to buyers, and with the central strategy of utilizing advice, convenience, and service, the next steps were to reenergize staff, raise morale, and enhance their skills through effective training to implement Joly's strategies. The Best Buy case also ties into the point about offensive versus defensive actions, which will be discussed in subsequent chapters.

Given the initial competitive situation where Best Buy's competitors attacked first with low price to draw away shoppers, Best Buy in this opening stage can be characterized as being the defender. Joly knew he would have to respond or forfeit his market position and watch a continuing drain on Best Buy's revenues.

Still in his defender's position, he scrutinized the opponent's offensive efforts, which were fully exposed to him. The competitors, at that point in time, could not know for sure if, when, or how Best Buy would respond to the price attack. For that single moment the psychological advantages existed with the defender, Best Buy, unless Joly decided to forfeit the market to the low-cost competitors and remain passive.

Joly did not remain passive. His defense aimed at blunting the competitors' primary strategy by matching their low prices. Then, he went on the offensive to regain the advantage by focusing on his combination of excellent advice/convenience/service to customers. In effect his turnaround strategy neutralized the rivals.

PreEmptive Campaigns against a Competitor to Blunt His Aggressive Actions before He Begins, Thereby Preventing Drawn-Out and Costly Market Warfare—The central issue in this type of campaign is to avoid long drawn-out campaigns that drain resources. Further, it has the debilitating effect of causing excessive tension and fatigue among personnel. The exception is where your strategy intentionally avoids conflict, with the purpose of exhausting the competitor.

Such an approach begins with a thorough competitor analysis, which might reveal a lack of resources and, thereby, an inability to hold out in a drawn-out campaign, or the analysis could uncover patterns of behavior to show that its leadership is unsteady and tends to make erratic moves. Simply waiting, doing nothing would be an anathema to such a manager. In still another instance, an impulsive manager may act with some ill-conceived and flawed plan.

Campaigns Tied to Obligatory Commitments Entered into through Joint-Venture Agreements—With the variety and prevalence of joint ventures and various forms of partnerships, temporary or permanent, there could be obligations to come to the defense of the partner. This is especially so where the reason for the alliance was to strengthen a market position by defending an exposed market niche.

Campaigns of Opportunity—These campaigns are usually outcomes of strategic business plans where expansion is a primary objective. Underlying such a goal usually requires preserving resources for use in upcoming opportunities. What follows is preparing skillful strategies that focus on discovering and exploiting a competitor's weaknesses, such as poor product and service performance, lapses in technology, problems with the supply chain, or inept leadership. A more comprehensive list of categories appears in Table 1.1.

Campaigns that Expand into Additional Market Niches Where There Are Dominant Competitors—Campaigns of this type rely on a two-pronged strategy: First, the use of segmentation to define and serve customers that match your criteria; second, the use of segmentation to locate the decisive point through which to concentrate resources against the vulnerabilities of your rival.

That point of concentration takes place where you and your competitors interact. That is, it takes place within market segments and their smaller

Identify the Root Causes That Trigger a Turnaround • 11

TABLE 1.1

Categories to Discover and Exploit Competitor's Weaknesses

Markets	Product or Service	Price	Communications	Supply Chain	Personnel	The Organization
Areas of opportunity by: • Market segments identified as emerging, neglected, or poorly served • Market weakness indicated by reputation, level of market penetration, or availability of sustainable services • Level of commitment to the long-term development of a market; level of investments in people, research, and technology • Ability to sustain long-term customer relationships	Areas of weakness by: • Quality • Features • Reliability • Packaging • After-sales services • Warranties • Returns policy • Level of technology • Extent of applications • Brand name awareness • Stage in product life cycle • Frequency of new product introductions	Price leader or price follower • Level of discounts, allowances, credit, and financing terms	Comparative weaknesses in: • Advertising commitments (business-to-business, business-to-consumer) • Sales force (selling skills, training, sales aids, incentives, compensation, motivation, market coverage) • Sales promotion (trade shows, webinars, contests, premiums, coupons, other) • Telemarketing and mobile communications • Internet by usage and application • Publicity • Market research and competitor intelligence	• Vulnerability of sales force directed at end-use customers and/or intermediaries along the supply chain • Market coverage related to gaps in market reach • Inventory control systems • Physical transportation • Support systems along supply chain	Level of weaknesses by: • Customer-driven orientation vs. product-driven mindset • Morale and group unity displayed by competitive spirit and motivation • Market and competitor awareness and ability to foresee trends • Overall capabilities to deal with threats • Overall experience, skills, and training by job level • Overall experience and knowledge of the business or industry • Level of understanding about strategy and tactics and ability to apply techniques to market opportunities and competitive threats	• Weaknesses in company culture as identified by its history and patterns of performance • Type of work environment and management support • Extent of internal communications • Managers' ability to react to market opportunities and threats • Managers' competence in planning and developing competitive strategies • Commitment to ongoing training and development of personnel • Financial resources and ability to sustain operations

subdivisions, niches. Consequently, your job in this type of campaign is to effectively define a decisive point through which to enter a market. The point is similar to the Xerox case cited above where the Japanese firms found a vacant and growing segment among small and midsize firms.

What does a decisive point or segment look like?

From a broad market perspective, China is a prime example of several segments or decisive points for any organization attempting to enter that vast market. For instance, take its enormous population of 1.3 billion, speaking more than 100 dialects. That alone makes China about as diversified as any single country can be from a marketing perspective.

It is with such immense diversity that segmentation plays a central role in developing a viable portfolio of opportunities. For instance, gateways open to what people eat, wear, and drive—from north to south, east to west, rich to poor, young to old, city to countryside. From a manager's position, then, China stands out as a superb example of the potential opportunities when using decisive points to penetrate a market.

Succeeding by focusing on decisive points or segments incorporates a four-step process:

1. Employ market intelligence and big data to define customer segments and evaluate competitors' vulnerabilities.
2. Conduct a comparative analysis to identify one or more decisive points, represented by a market segment that can be characterized as neglected, poorly served, or emerging, or by the competitors' weaknesses in such areas as service, product performance, logistics, leadership, and the like (again, refer to Table 1.1).
3. Pinpoint a segment for initial entry; once secured, then systematically roll out into additional market segments, still using comparative analysis and market intelligence and identifying competitors' vulnerabilities.
4. Customize products and services to those segments with unique and definable benefits.

Thus, employing segmentation and focusing on a decisive point means splitting the overall market into smaller submarkets or niches where you enjoy measurable and substantial advantages to serve your customers and win over competitors.

Campaigns to Solidify an Existing Market Position and Make It More Defensible—Referring again to the Xerox case, the company has to be admired for virtually reinventing and redefining an industry with its innovative technology. Xerox correctly focused on the likely early adopters—large organizations. It further established a brand awareness that has been recognized worldwide for decades with the often used phrase: "Make a Xerox copy."

What, then, is the issue? It didn't cover its flanks. Xerox management failed to adequately solidify its market position and make it defensible against the inevitability of other innovative companies aiming, not just as a point of entry into the copier market through small and midsize firms, but eventually to use it as a foothold to encroach directly on to Xerox's turf.

A speculative viewpoint could argue that there wasn't an urgent need to maintain a full 360-degree awareness during the 1970s to signal an impending competitive crisis coming from Japanese firms entering North America through an undefended segment. Or were there some early signs of a major effort building? Were there reports coming in from the field that were ignored? Didn't the business press play up the action through major stories that perhaps were not read or taken seriously by Xerox managers?

Further conjecture could speculate that the structure of the organization didn't permit the flow of vital information through the internal layers of Xerox management, and the leadership at the time did not see a looming competitive emergency. Whatever the actual causes, a crisis did occur. Xerox survived. A turnaround plan was eventually put in place for the hard fight back to rebuild a viable market position and make it more defensible.

Campaigns into New Markets or New Businesses in Support of Long-Term Strategic Objectives Related to Growth and Expansion—The objectives set forth in a strategic business plan at some point have to translate to action. Typically, objectives include market entry with new products. Such campaigns require intelligence about the needs of the end-use customers, the logistics to reach them, the decisive point through which to enter, the kind of strategies to overcome resisting competitors, and a poststrategy to defend what has been achieved.

Campaigns by an Aggressive Competitor When It Attempts to Weaken the Defender's Resistance—As an extension of the above type of campaign, this one consists of a two-pronged event: (1) introduce a tangible product or service, and (2) use a psychological subterfuge that induces fear to weaken the defender's resistance.

Such is the case of an air conditioner manufacturer that intentionally leaked announcements to the press and at trade shows that its breakthrough technology would shake up the industry. The information was purported to be a major advance in air conditioning innovation where products have essentially remained unchanged for decades. Further, the company announced that the first delivery of its new product would take place in eight months, when, in fact, the product launch was planned for three months to achieve a psychological surprise among competitors.

Achieving a psychological advantage creates an unbalancing effect in the rival manager's mind, whereby he or she vacillates in indecision. The further intent is to disorient and unbalance the competing manager into wasting time and making costly and irreversible mistakes. Using distractions and false moves make it appear that you are launching your effort directly at the competitor's strengths. Whereas, your true purpose is to target his vulnerabilities.

Moreover, the purpose is to convince your competitor that continuing aggressive efforts would be too costly, with little or no chance of justifying the expenditures of people, money, and materials. Consequently, fighting in the marketplace is not your intention. Rather, your aim is *possession*. That is, your purpose is to hold a long-term position in a target market, as gauged by attaining a market share objective, securing a position on the supply train, reaching a profitability goal, or similar metrics.

Campaigns That Are Limited-Term Tactical Moves versus Long-Term Decisive Campaigns with Broader Strategic Objectives—The central issue here is to determine if the limited-term campaign is an impulsive one of running after some Monday-morning headline, which consumes resources and diverts management's attention. Or is it one with limited objectives, but connects with the overall strategic direction of the firm or business unit?

For instance, if the limited-aim campaign is conducted to shore up a vulnerable part of an exposed market segment that would give an opening to an enterprising competitor, then there is justification. Again, refer to the circumstances surrounding the Xerox case to sense the marketplace realities.

Campaigns That Reverse Declining Sales and Reenergize a Demoralized Company's Personnel—This type of campaign is similar to the Best Buy case cited earlier. You also may see a connection with one or more of the above types of campaigns, because reversing declining sales and reenergizing personnel have universal commonalities.

In another case, LEGO (the Danish toymaker) at one point was on the brink, losing $1 million a day and posting record deficits. That was a decade ago. A turnaround plan subsequently was put in place. Discarding the traditional tools of research, such as focus groups to determine new product lines, managers employed what they defined as an anthropological approach of understanding human behavior. Using a consulting firm that relied on such techniques, anthro teams fanned out across Los Angeles, New York, Chicago, Munich, and Hamburg asking kids to tell them stories, interviewing parents, and making photo diaries. As a result of that undertaking, today LEGO is a vibrant company with near all-time high sales.

Campaigns Intended to Make the Confrontation More Costly for the Rival to Continue Operations, Thereby Neutralizing the Competitor and Rendering Him Harmless—These campaigns contain physical and psychological considerations when devising any strategy. Fighting is not the object of the campaign, as it can cause excessive expenditures of resources; rather, it is possession of the marketplace as the true objective. In such a case, any strategy that aims to wear out the rival by making further action overly costly to him is acceptable. Making the campaign more costly for the competitor was included in the comments above on preemptive campaigns.

Campaigns Initiated by Upward Pressures from Junior-Level Managers to Take on More Ambitious Objectives—As executives rise in rank, there is evidence to show that the brashness and aggressiveness they used to climb the career ladder had often leveled off. A deeper awareness of the magnitude of the responsibilities takes over. An internal sensitivity as to how a faulty decision can seriously affect the human welfare and material resources of the organization becomes top of mind. As a result, the same flairs and impatience for action they displayed in earlier days now creates pressure from the up-and-coming managers.

Thus, such campaigns take a good deal of disciplined judgment to decide whether to risk a campaign or turndown and possibly turn off junior-level managers.[*] Nonetheless, a judgment call has to be made. Of course, there

---

[*] See Table 1.2 (Conditions when conflict should be avoided).

is always the possibility that the junior manager may be correct and there is an overly cautious approach at the senior level.

PHYSICAL AND PSYCHOLOGICAL CHARACTERISTICS OF A COMPETITIVE CONFLICT

Having looked at the primary causes of a competitive conflict and discussed the various types of campaigns, let's dig deeper into the following primary characteristics of a competitive conflict.

Conflicts Do Not Break Out Unexpectedly

For the most part, competitive confrontations are not likely to break out without advanced signs from the marketplace. Nor should they come about as any surprise, assuming, of course, the market in general is carefully watched and, in particular, competitors are closely observed.[*]

As will be discussed in greater detail further on in this and subsequent chapters, entering into a major encounter with a competitor should be the consequence of a business plan's strategic direction, as well as from the plan's objectives. Thus the cliché, *choose the battles you wish to fight*, has a pragmatic meaning in the context of planning for a long-term turnaround.

Consequently, encounters should not be carried out randomly. However, where you are pushed to the wall and must protect your market position, such campaigns are inevitable. However, deciding on how the confrontation would be handled should still keep the strategic picture in mind.

Such was the case of *Yahoo*. Within one year of Marissa Mayer taking over as CEO in 2012, she moved rapidly with sweeping changes, replacing key executives, changing internal work rules, making a string of 22 acquisitions, overhauling the company's culture, and transforming the company's strategic direction into a media company. The central aims behind Yahoo's calculated moves, which committed substantial expenditures of

---

[*] This point applies to movements of major size and magnitude, which should be visible to any alert manager; whereas surprise is more likely at the local level, as with a competitor launching a sudden wave of targeted promotions, introducing short-term price discounts, and the like. Here, too, flexibility and anticipation of such events must be expected and a rapid response set in motion.

financial and material resources, were to make a sweeping turnaround and reverse declining revenues, rebuild market presence, and energize Yahoo's personnel for expansive moves.

At the same time, however, Mayer was sending clear-cut, highly publicized messages to vigilant competitors that signaled offensive actions were on the way. Thus, the stage was set for inevitable confrontations against equally committed competitors who were not expected to sit quietly on the sidelines and see penetration of their respective market areas.

Conflicts Require Neutralizing the Competitor

The purpose of weakening the rival's capabilities is to get it to the point where the competitor cannot, or will not, take any further action against you. Expressed another way, the intent is to cancel the competitor's advantage, or else to diffuse it, so that the primary strategies are neutralized. The key idea is, if you don't take action quickly to minimize the competitor's capabilities, the possibility exists that he will attempt to take action and cancel out your lead.

The tools to neutralize a competitor can be physical or psychological and can take place in numerous ways, from using innovative technology to weaken the competitor to employing fear to demoralize its personnel. Such was the case of Apple versus BlackBerry.

The Canadian company introduced its BlackBerry prototype in 1997, with a dominant feature of permitting users to get their emails on the move. Soon after launch, an explosive period of growth began. The device developed an iconic status whereby owners of a BlackBerry acquired prominence.

The prestige image was perpetuated as the likes of Barack Obama, Hillary Clinton, Oprah Winfrey, and high-profile Hollywood celebrities were proudly seen using their BlackBerrys. By the end of 2004, the Waterloo, Ontario, company sold its product line through 80 carriers in 40 countries, and boasted more than two million subscribers.

Then, in 2007, an audacious outsider appeared in the marketplace and challenged BlackBerry: Apple's iPhone. Rather than recognize the new product as a potential competitive threat, BlackBerry's over-confident leaders publicly belittled Apple's device. At the same time, internal flaws began to appear within the company as research took a backseat and complacent attitudes prevailed among numerous executives that there was no need for fear from iPhone as a threatening challenger.

In 2008, the advanced iPhone 3G and the first Android phones went on sale. BlackBerry responded with its new product introductions, but they never fully equaled the technology and features of the competing products. In all, the internal weaknesses in product development and neglect of marketing at BlackBerry began to show more acutely as key personnel began leaving the company.

Then, in 2010 the decisive blow came when Apple introduced its iPad, one of the most successful consumer products ever. BlackBerry's attempt at a turnaround to counter the move by introducing its PlayBook tablet a year later resulted in a dismal failure. Apple totally neutralized BlackBerry's advantages.*

Whereas technology might have a decisive impact in a competitive conflict, as in the Apple versus BlackBerry case, most types of confrontations require a combination of tools applied in a series of campaigns to achieve any serious impact.

There are two central points to keep in mind in selecting the tools. First, your approach is to match your capabilities and areas of strength against those of your competitor, so that you can clearly define the vulnerabilities of your rival (see Table 1.1).

Second, as you develop the selection criteria, keep in mind the psychological impact any combination of tools would have on the competitor's personnel and their power of resistance. Dispirited, discouraged, and unmotivated individuals can be the critical psychological factors in the outcome of a conflict.

Therefore, carefully observe your opponent's leadership. Monitor behavior over a period of time and in a variety of marketplace events. For instance, is the competitor likely to react impulsively in a fight-or-flee mode? Or is behavior more deliberate with a wait-and-see approach that vacillates between active or passive? And what types of actions are likely to trigger an aggressive response?

From such observations, you can develop your strategies and make reasonably accurate estimates of the types and levels of resources needed in winning a confrontation. These assumptions are made, however, with the understanding that absolutes about behavior are not totally accurate. An opponent's power of resistance is not altogether quantifiable and the human mind cannot be ruled by logic.

---

* As of this writing, another BlackBerry turnaround is in the making.

On the other hand, behavior is not a totally unknown factor. It is still quite reasonable to forecast patterns of conduct, performance, and comportment based on empirical evidence of what was done yesterday, providing you have some insight about the culture of the organization and the existing leadership style.

Thus, just as you can gauge the other to a large extent by what the competitor is and does, so he can judge you accordingly by similar astute observations. And by accepting such feasibility, you also can assume that conflict should not break out wholly unexpectedly, nor can it be spread instantaneously.

Conflicts Are Not Isolated Events

Rarely is a successful confrontation with a rival ever achieved in a single decisive campaign. The reality is that successive confrontations are the norm spread over long periods of time. Numerous factors underlie this principle, which are evident in the following example.

Samsung, the South Korean behemoth, recognized the importance of staying alert and preparing for the inescapable and ongoing pressures of competition. Although the company achieved outstanding results in 2013, all employees still received an email message from CEO Lee Kun-hee: "You must do better!"

What dramatizes his to-the-point exhortation is that Samsung had reached extraordinary levels with sales about equal to those of Microsoft, Google, Amazon, and Facebook combined. The company shipped 215 million smartphones, about 40 percent of the worldwide total, with predictions of 350 million the following year. And it is valued as one of the most profitable companies in the world, with the Samsung brand ranked as the eighth most-valuable worldwide.

So why all the urgency? Several issues appear to have been embedded in Lee's mind when he sent the email.

First, there was the legitimate concern with the inherent danger of complacency, which reflected in his follow-on commanding statement: "As we move forward ... we must start anew to reach loftier goals and ideals."

Second, he was worried about what is called the fast-follower problem, or, in marketing parlance, short life cycles followed by rapid competitive product introductions. Although smartphones have been the major driver of Samsung's growth in recent years, management internalized the fleeting nature of mobile phone leaders.

All Samsung managers were well aware of brands that peaked and then plunged: Motorola, Ericsson, HTC, Nokia, and the previously noted, BlackBerry. Then, there were the low-price smartphones hitting the market from ambitious Chinese brands, such as Xiaomi, that were anxious to export internationally.

Third, Lee knew that only developing product line extensions were no longer enough to stay on top, he needed to push employees to think more boldly with the all-inclusive approach of integrating every aspect of a new product development, including design, software, and hardware. That strategy was demonstrated by Samsung being first in the market with its Galaxy Gear smartwatch. Lee internalized the time-tested wisdom that competitive conflict is not won by a single decisive event.

Campaigns Cannot Be Interrupted

As pointed out in the Samsung case, competitive activity is ongoing, which means that whether an organization wins or loses in the initial confrontation, activities cannot be interrupted. That is, the ultimate outcome of any single campaign is never to be regarded as final. Rather, it must be considered as a transient event, for which either side has to ready itself for successive clashes until actual success is achieved. Also, any interruption gives the loser time to recoup with fresh resources and usually fresh strategies.

How, then, should success be measured? Measures include (1) neutralizing the rival's capabilities, (2) causing the contender to exit the market voluntarily or through bankruptcy, or (3) learning to adopt a live-and-let-live policy where there is mutual interest among competitors for the long-term growth and prosperity of the market.

The significant point: Avoiding complacency, acting with boldness, and motivating employees to win competitive conflicts are other ways of expressing Lee Kun-hee's intention of "You must do better."

Yet, still other issues are embedded in this point of not interrupting campaigns. First is the ongoing training of individuals, second is the positioning of forces for defense or offense.

As for training, it should be evident at this point that raising the skill levels of individuals is an open-ended necessity. Yet, it still is given haphazard acceptance in some organizations where misguided, redundant, or underfunded programs exist. Worse yet, the training doesn't relate to the strategic direction and objectives of the organization.

Training, within the context of developing a turnaround plan, means equipping your staff with skills related to crafting competitive strategies. These, in turn, would convert into distinctive campaigns to deal with the types of competitive encounters mentioned earlier. This point takes on special meaning when individuals are often the deciding issue, where all other competitive factors are about equal.

Further, training sessions should aim at creating a state of readiness. The key word, *readiness,* applies to taking on an opponent that is going after your company and your market position. Specifically, training programs should relate to the following guidelines:

- Train individuals to develop a competent strategy plan that leads to any of the three measures of success listed above. Components of that training should include making certain that key personnel at every level internalize the strategic direction of the business plan, writing objectives that complement the long-term corporate or business unit vision, preparing strategies to achieve the objectives, and proposing products and services to outdistance rivals and maintain a competitive lead.
- Train all levels of staff to the vital importance of acquiring ongoing competitor intelligence (CI). This is the key component to developing and implementing competitive strategies.
- The central idea behind CI is that if you know your rival's plans, you are able to monitor his moves. Then you can figure which strategies will likely succeed and avoid those with little chance of realization. You, thereby, gain the advantage of knowing where your competitor's strengths are formidable and where weak.*
- Initiate specialized training that focuses on techniques to neutralize a competitor. That includes gaining the input from a variety of viewpoints and sorting the information into a coherent and actionable plan.

---

* The ancient Chinese strategist, Sun Tzu, summed up the purpose of competitor intelligence 2,500 years ago. To get the full benefit of his meaning, substitute the word *enemy* with competitor. "*Know the enemy* (competitor) *and know yourself; in a hundred battles you will never be in peril. When you are ignorant of the enemy* (competitor), *but know yourself, your chances of winning or losing are equal. If ignorant both of your enemy* (competitor) *and of yourself, you are certain in every battle to be in peril.*"

- Establish interactive communications within your organization that encourage innovative thinking among all levels of personnel. That means sharing forward-looking plans and fostering full participation.

One of the overall outcomes of effective training is that it creates an environment of trust and understanding, whereby subordinates are encouraged to seize the initiative and act with a sense of purpose and loyalty. Therefore, a parallel dynamic that embraces training is leadership, which centers on motivating individuals to bold and decisive action.

As noted above, an all-important role of training is to sensitize individuals to the strategic direction of the organization or business unit. The essential meaning underlying this point is to maximize the efficiency of the staff and to concentrate resources to achieve the planned objectives. Doing so guards against running off after every diversion that comes along, which can have the damaging effect of distracting management into head-on collisions against competitors *without* adequate preparations.

Leadership, then, is about responsibility, accountability, and achieving objectives. Effective leaders inspire their people, organize actions, develop strategies, and respond to market and competitive uncertainty with speed and effectiveness. Above all, leaders act to win customers, to win market share, to win a long-term profitable position in a marketplace, and to win a competitive encounter before a rival can do excessive damage. If they lose, their organizations and those they manage suffer.

FACTORS THAT CAN BRING A CAMPAIGN TO A STANDSTILL

Notwithstanding all the advanced techniques, processes, and guidelines to prepare for conflict, there is an unsettling reality that numerous factors can bring your turnaround campaign to a standstill. The most blatant one is poor intelligence of the competitive situation, which can cause anxiety, hesitation, and skepticism to enter your mind. And, if those thoughts linger and are observed by those you manage, they can penetrate their minds and create negative attitudes that can affect the outcome of your campaigns.

For instance, how complete is your knowledge about your competitor's intentions? Are they local efforts or major expansive moves into several markets? How many resources will the competitor likely throw against your market position? Does the competitor really have a knock-out strategy that can do substantial damage to you, or is it a ruse? How good are your defensive and offensive capabilities?

As for defense versus offense, the defense is a stronger form of battling it out in a confrontation, as long as the defender understands the two-step process: first, gathering intelligence and developing a counterstrategy, and second, implementing the strategy. Expressed another way, the waiting and watching defensive period is only a prelude to an offensive strategy.

Best Buy, cited earlier, is an example where defense is superior to attack. In other instances, where there is overwhelming marketing power, breakthrough technologies, or other totally new innovations used by the attacker with which the defender cannot cope, then any chance of the defender launching an effective response is problematic.

Still, organizations can find remedies, such as finding niches to continue operations or forming alliances with technology-rich firms. (More on defensive versus offensive actions in a later chapter.)

Another reality is the lack of full information about your own firm's situation. Are you fully aware of the physical capabilities that define your organization's strengths and weaknesses? How strong are your company's financial resources? Where does your organization stand on known and unknown technologies that may impact your campaigns?

From still another viewpoint, are there cultural innuendos that at times seem intangible, yet are prevalent and influence the decision-making process? Are there latent forces that permeate the organization and appear to drive the attitudes, which psychologically affect the morale of managers and create friction among the rank-and-file? Are they powerful enough to form positive or negative mindsets and behavioral reactions to competitive stress?

The implication is that if you don't have substantive answers and you are hard pressed to gather all such knowledge, then it is possible that your turnaround plans can grind to a standstill. Yet, any of those factors can be the platform on which to develop a strategy to launch a campaign or defend an existing market segment.

Thus, the importance of internal company information and the vital need for competitive intelligence remain a priority of the highest magnitude. And, even with all the sophisticated techniques and analytics available, it is still imperfect when it comes to the fine points of understanding the workings of the competitor's mind. Therefore, lack of knowledge of these factors can suspend action on your campaigns.

There is also the potential danger that any serious errors in assembling the information can lead to the flawed conclusion that the initiative may lie with the competitor, when, in fact, it could be with you. Such an error can result in an ill-timed campaign or an unfounded delay that also can suspend action.

Then, there is a parallel condition where human nature becomes a key issue. It is the inclination for individuals to estimate the opponent's strength as too high rather than too low. Such miscalculation is also a factor in delaying the progress of a campaign.

Notwithstanding that these multifaceted factors do exist, the remedy is to pay close attention to competitor intelligence. Once assembled into what you believe is reliable data, you still have to rely on your best judgment, training, experience, and intuition to move forward.

The following case of Starbucks' entry into supermarkets illustrates some of the above points. The company used a strategy of setting-up dedicated aisles, or Signature Aisles, to display its coffee products and new noncoffee offerings, such as yogurt products, bakery items, juices, and healthy snacks.

What exactly did Starbucks get into? What type of intelligence would have helped in the launch? Most supermarket aisles were typically controlled by big-name producers that diligently aim at maintaining a dominate presence on store shelves for their respective products. Yet, even the likes of Kellogg, Heinz, and Campbell with their sophisticated intelligence-gathering methods and their high levels of supermarket experience have had to pick up the remains of failed product launches.

Certainly, Starbucks did its requisite market research. In the end, however, there were still substantial unknowns that could have brought the major move into supermarkets to a standstill. Yet it seems from all indications that the experience and intuitive feelings of CEO Howard Schultz became the driving force to move forward.

Then, there is a group of diverse influences that can bring a campaign to a standstill, any of which can affect implementing a turnaround plan. These include the staff expending excessive physical energy, with the eventual impact fatigue has on job performance; personnel exhibiting confusion as to the direction of the organization; employees showing loss of morale, along with their concerns for job security, prospects for personal

TABLE 1.2

Conditions When Conflict Should Be Avoided

1. Personnel are not ready for a competitive conflict, or they are discouraged by previous failures and morale has not sufficiently improved.
2. Superiority of resources is viewed as the deciding factor in the outcome of a sustained campaign, and the defending organization is materially weaker with little to no chance of obtaining additional means.
3. The focus of the conflict is in an area or market that is totally unacceptable to the capabilities and background of the organization.
4. The organization would best benefit by the passing of time instead of entering a conflict, such as in the case of expecting aid through an impending merger or partnership.
5. The competitor is expected to voluntarily exit a market due to its own lack of resources to carry on a prolonged campaign.
6. It is the wrong time to enter a conflict due to disruptions in the industry, signs of shifting consumer trends, or unstable economic conditions.
7. The prospects of winning a campaign represents too huge a gamble or the potential loss of resources and market share are too severe; in contrast, should the competitor win, the benefits would reach far beyond the scope of the campaign itself.

---

growth, and working conditions. Further, lack of confidence in the leadership can create a dispirited environment.

At the same time, there are the largely uncontrollable events and unsettling challenges of individuals' personal lives outside the job that can create unexpected forms of resistance, which could jeopardize a plan.

Consequently, there are times when it may be appropriate to avoid conflict altogether, which also can be regarded as a turnaround. Depending on circumstances, that could mean going completely on the defensive to protect what you already hold. That is, you feel confident that you have sufficient strength to prevent a competitor from dislodging you. These conditions, which should be given your attention, are summarized in Table 1.2.

Competitive Conflicts Contain Elements of Chance

What follows, in part, from the Starbucks case is that myriad influences and uncontrollable events can suspend action. In turn, those conditions introduce the element of chance, which some may characterize as luck. Thus, developing a turnaround plan and the related strategies comprise many unknowns, which hover within the ever-present cloud of unintended possibilities and probabilities.

Consequently, given the total framework of competitive conflict, turnaround plans, and controllable and uncontrollable elements, what summarizing conclusions can be drawn thus far?

First, a crisis most often can be identified with preexisting causes. Therefore, it is important to isolate the fundamental reasons that precipitated a need and take remedial action.

Second, a dominant guideline when shaping a turnaround plan is that the resulting campaigns are firmly connected to the organization's strategic direction.

Third, the successful implementation of a turnaround is linked to a strong corporate culture that embraces flexibility, determination, and readiness. Such attributes rely on individuals fortified with applicable skills and energized through high morale.

Finally, within the framework of the 13 types of competitive campaigns mentioned earlier, there are physical, psychological, and the vagaries of chance that can bring a campaign to a standstill. Now move on to Chapter 2: Prepare the Organization for a Turnaround.

2
Prepare the Organization for a Turnaround

Chapter Objectives

Be able to

1. implement a system of top to bottom planning to integrate the thinking of diverse groups of individuals;
2. align the organization's culture with the turnaround planning process;
3. link morale to the turnaround strategy through a unified staff; and
4. strengthen relationships between leader and staff to develop a turnaround plan.

INTRODUCTION

The previous chapter highlighted the categories of conflicts. Within that framework, it was clearly stated that a competitive crisis is not an isolated incident. Rather, it is inextricably tied to some preexisting internal conditions or unusual external events that were left to fester within an organization over an extended period of time.

The most vivid example was the General Motors' ignition switch problem that was suppressed over a period of years and resurfaced in 2014. The incident was headlined by a well-known business publication as: "GM still hasn't fixed its problematic culture in the years since bankruptcy."

In contrast, where a crisis is averted or managed with speed and efficiency, most often the result can be traced to an alert leadership, a well-developed strategic business plan, and trained and motivated personnel. Further, it is backed by creative strategies supported by excellent competitive intelligence that aims to neutralize the rival's capabilities.

Within those views, two dimensions characterize a crisis: the physical and the psychological.

THE PHYSICAL DIMENSION

In the Best Buy case, cited in Chapter 1, the initial defensive move of lowering prices to cancel out the competitors' pricing advantage and then following with an offensive attack spearheaded by a superior advice/service/convenience strategy were physical acts.

Addressing the physical dimension is an essential issue in preparing for a turnaround. It reflects on how the firm is organized for efficient implementation of the plan. That means simplifying the system of command and control, maximizing communications and decision making, and eliminating the organizational layers that form physical barriers from the field to top-level executives, which prevent the plan from moving forward with consistency and speed.

In a small and somewhat less complex organization where a president is normally at the helm, he or she is in a unique position to control both policy making and execution of plans. In the larger multiproduct firms with more people, products, and additional levels of authority, results may fall victim to cumbersome processes and procedures.

Individuals in the field often feel that there are obstructions in the decision-making process for moving into new markets. Missed opportunities are common, and "go" decisions get stuck for reasons other than the competition.

Your own experience may well support the obvious inference that an organization with many levels in its decision-making process cannot operate with speed. This situation exists because each link in the managerial chain carries four drawbacks:

1. Loss of time in getting information back
2. Loss of time in sending orders forward
3. Lack of full knowledge of the situation by senior management

4. Reduction of the top executive's personal involvement in key issues that affect the availability of resources

Therefore, for greater efficiency and speed, it is advisable to reduce the chain of command. The fewer the intermediate levels, the more dynamic the operation. The result is improved speed and increased flexibility.

A more flexible organization can achieve greater market penetration because it has the capacity to adjust to varying circumstances more rapidly. Thereby, it can concentrate at the decisive point before its competitors have a chance to respond.

Flexibility also permits a quick response to an impending crisis, especially where the need is incorporated into a strategy team's responsibilities (Table 2.1).

THE PSYCHOLOGICAL DIMENSION

Once it appeared to the discerning eye that Best Buy's strategy would win, the behavioral aspect of its personnel would have revealed a spirited, elated, and a nothing-can-stop-us attitude.

In contrast, the competitor's personnel would have seemed played out, unable to carry on the conflict to a decisive end. Their behavior would have been seen as disheartened, lacking in enthusiasm, with no will to carry on. In effect, their performance would take a sharp decrease in overall effectiveness. Expressed another way, the competitor's capabilities would have been neutralized.

Two major forces shape the foundation of this psychological dimension: organizational culture and morale.

Organizational Culture

Culture shapes the behavior of personnel and reflects their feelings and actions. It encompasses the organization's values, visions, norms, working language, systems, symbols, beliefs, habits, and its history. Together, they form the DNA that *is* the organization. Culture is what makes the organization a living, working entity.

In practice, organizational culture foretells the manner in which the company enters a market and implements its strategies. It guides the

TABLE 2.1

Duties and Responsibilities of a Turnaround Strategy Team

Functions include:
- Define the root causes of the competitive crisis.
- Develop a collaborative approach to developing a turnaround business plan.
- Identify changes in the industry, economic issues affecting the market, behavioral changes among customers, and specific competitive threats.
- Select decisive points where the competitor is most vulnerable.
- Develop short- and long-term objectives and strategies.
- Prepare product, market, supply chain, and quality plans to implement competitive strategies.
- Maintain an early-warning system to detect internal and external conditions that would signal an impending crisis.
- Align the turnaround strategy plan with the organization's culture.

Responsibilities include:
- Create and recommend new or additional products and services that would be included as part of the turnaround plan.
- Approve all alterations or modifications of a major nature.
- Serve as a formal communications channel from the market back to internal departments.
- Plan and implement strategies that utilize competitor intelligence to determine defensive or offensive campaigns that can neutralize the competitor's efforts.
- Develop turnaround programs to improve market position and profitability.
- Identify product and service opportunities in light of changing consumer buying patterns.
- Coordinate efforts with various corporate functions that take a collaborative approach to developing short- and long-term objectives.
- Maintain ongoing communications to encourage interdivisional exchanges of new market or product opportunities.

selection of employees, influences management styles, determines the timeliness of adopting a technology, and influences the organization's structure. Culture impacts the way individuals and groups interact with each other, with intermediaries, and end-use customers.

Such is the case of Monsanto, which prides itself on its collaborative culture. The entire company embraces the mission that seeks to combine the abilities of executives, scientists, employees, farmers, and other stakeholders in producing more while conserving water, energy, and other natural resources. As Monsanto's president, Brett D. Begemann, expresses its culture: "We work together more like a family than a corporation."

Implementing such a culture means steering unwavering attention on employees. That means talking to them, listening to them, taking a risk on their ideas, and accepting an occasional failure. It's about nurturing their experiences and sharing knowledge with them in a proactive way.

By means of organizational communications where stories, rituals, and symbols are used, employees through the ranks share meanings. And, through the existing and evolving electronic channels, core messages are transmitted in such areas as new product developments, industry changes, organizational and personnel achievements, and the like. In doing so, it also eases the onset of negative subcultures within business units that may be contrary to the strategic direction of the firm.

The overall value of culture is that it helps an organization survive and flourish. If the culture is valuable, then it holds the potential for generating sustained competitive advantages. Additionally, values and practices are learned through socialization at the workplace. That is, the work environment reinforces culture on a daily basis by encouraging employees to exercise cultural values.

Then, there is the situation where organizational culture can have a negative effect as in the dramatic case of J.C. Penney and its attempt to implement a turnaround. Former CEO Ron Johnson began the process with his personal vision of creating a new type of retail establishment as a pathway to emerge from Penney's lackluster performance spanning several years.

The central idea, however, was to totally transform the 110-year-old retailer from its roots in Middle America to an upscale, youth-oriented store that is considered cool. However, from all evidence, such ideas and means of implementing change didn't originate with a documented strategic plan; rather, many occurred through evolving strategies.

The process consisted of totally upending Penney's culture, practices, and policies from top to bottom. It meant a number of extreme and rapid changes, such as installing a completely new and radical pricing and merchandising model that had never been used before, remaking the stores' interiors, removing standard merchandise that would have sustained a cash flow through the transition, and mixing new personnel with older management—the latter causing personality clashes from the onset.

The bitterness between the two management groups was exacerbated when some newly hired executives refused to move to Penney's headquarters in Plano, Texas. Instead, such key individuals from president, chief operating officer (COO), to Human Resources remained in other cities and jetted in weekly. That routine alienated them from day-to-day issues and the generally accepted practices of senior leadership, especially through a highly sensitive period of massive transition, where visibility would have been the more effective approach.

Then there was Johnson's attempt to make sweeping changes by slashing into Penney's deep-rooted culture with an extravagant and costly approach to change mindsets through elaborately staged events. It ended with a damaging effect on the 159,000 employees, most of whom hadn't completely bought in to Johnson's turnaround strategy.

The result was that the attempted turnaround failed through excessive draining of cash, the inability to gain the enthusiastic support of personnel, the alienation of Penney's mainline customers, and the inability to attract new ones.* Johnson resigned, the previous CEO Mike Ullman was reinstated, the company ran an apology ad for misleading the customers, and a fresh new turnaround began.

From the Penney case, several lessons emerge as guidelines to increase the chances of successfully activating a cultural change within an environment of competitive conflict during a turnaround period.

Seek Maximum Input from All Levels of Employees

Serious attempts at soliciting input from seasoned Penney executives rarely existed during the turnaround attempts. To initiate new ideas that lead to new products, evolving markets, or new businesses requires a cultural sensibility that retains an open mind and avoids the idea-killing verbiage: *We've tried that.* To that end, utilize a cross-functional team to exchange ideas, foster cohesiveness and cooperation (Table 2.1).

General Electric is an exemplary company that illustrates the interplay of culture, staff involvement, competitive spirit, and creativity. "I'm intense about our competition, but I'm more concerned about our culture and our people," declared CEO Jeffrey Immelt.

He defined his concern when he admitted to two fears: first, that GE would become boring, and, second, that his top people might act out of fear. He meant that some executives would shy away from taking the essential risks to propel the company forward.

However, Immelt knew that in a slow-growth domestic economy and a volatile global marketplace, he had few alternatives other than to go on

---

* The comments about J.C. Penney are not intended to evaluate the retailing strategies. Actually, a few innovations worked quite well. Rather, the intent is to dramatize the underlying flaws in preparing a turnaround plan and implementing it.

the offensive and push boldly into new products, services, and markets. Taking a less risky approach would mean a backward slide from which it would be difficult to recover.

How, then, is a corporate culture overhauled, where its beliefs, practices, and creativity have been the hallmark of excellence, and where the company remains the envy of executives from other high-profile companies?

The following three steps summarize GE's cultural preparedness:

1. Compensation is linked to new ideas and customer satisfaction, with less emphasis on bottom-line results. It is based on managers' abilities to improve customer service, generate cash growth, and boost sales, instead of simply meeting profitability targets. Top executives hold phone meetings every month, meet each quarter to discuss growth strategies, and evaluate new ideas.
2. Executives must go after businesses that extend the boundaries of GE. More than just giving lip service to the order, they must submit at least three "Imagination Breakthrough" proposals per year for evaluation and possible funding. The criteria for submitting the proposals must include taking GE into new lines of businesses, geographic areas, or customer groups.
   However difficult the shift in mindset toward embracing risk, Immelt supports the effort by fine-tuning all the internal operating systems to make it happen. In other words, the risks are shared and are now culturally indigenous to the firm.
3. Executives are rotated less often, and more outsiders are brought in as industry experts instead of professional managers. That's a big departure from GE's promote-from-within tradition. Immelt pushes hard for a more global workforce that reflects the markets in which GE operates. He also encourages GE's homegrown managers to become experts in their industries rather than just experts in managing.

Stay on the Offensive

Although it is difficult to turn employees' mindsets from an impending crisis to energetic optimism, nonetheless it needs to be achieved if a turnaround plan is to work. That calls for a measure of boldness, rather than excessive caution. Other factors being somewhat equal, all evidence indicates that when boldness meets caution, boldness wins. Undue caution is countless times worse than excessive courage.

Looking at the history of business, enterprises accented with determination and purpose rather than by impulse, lead more often than not to successful performance. Staying on the offensive and using boldness have a powerful psychological impact on the employees. Whereas the use of excessive caution is handicapped by a loss of stability, initiative, and momentum. Therefore, when in doubt, some action is better than no action. (More detail on the value of bold action in Chapter 6.)

How far and fast you can move to the offensive depends on how well you can reinforce such a mindset among your employees. You need to feel a level of confidence that they are skilled, feel supported by management, comprehend the fundamentals of the turnaround plan, and fully understand the nature of the risks. In all, your task is to maintain a high level of morale by following these guidelines:

- Adapt to a flexible competitive environment. Flexibility is a singular characteristic that must be upheld. As such, it is an imperative for operating successfully, especially within the ambiguities of a competitive conflict.
- Maintain an outward display of resolute calmness and unshakable confidence. It fosters team solidarity and inspires a culture of innovation and creative thinking.
- Encourage an entrepreneurial mindset and promote employee involvement for a turnaround.
- Urge individuals to think about new markets, products, and services with the potential for establishing new revenue streams.
- Inform individuals of new company initiatives, which would prepare them to handle unexpected threats, or to react to time-sensitive market opportunities.
- Inspire individuals to win, which has a positive psychological impact on employee behavior and morale. Also, look out for signs of apathy or discouragement that stall momentum.
- Initiate skill development programs to instill discipline, and elevate employees' self-confidence, thereby preventing them from caving in under tough competition.

Act as an Aggressive Competitor

Similar in concept to the above guideline of staying on the offensive, this combative mindset helps in discovering your firm's comparative advantage, which is essential in developing a turnaround strategy. You, thereby, develop a sensitivity to look critically at strengths and weaknesses in products, services, logistics, and overall organizational structure. Further, it provides insight by examining relationships with suppliers, intermediaries, and customers along the entire supply chain.

The process exposes strong points and vulnerable areas in technology, manufacturing, human resources, and capital resources, as well as highlighting other areas that might endanger the firm to further competitive attacks. As important, it unmasks sensitive information on employee behavior and suggests clues on how to implement a turnaround.

Build a Strong Market Position

The aim is to create a defensible market position from which competitors cannot easily dislodge you. To that extent, create brand equity and brand recognition. All efforts should be directed toward mounting a long-term positive image for the firm.

Therefore, customer satisfaction and long-term customer relationships remain the enduring principles. This guideline is not in opposition to the previous one; stay on the offensive. That is, all defense strategies have an offensive component, which will be discussed in subsequent chapters.

Stay Close to Evolving Technology

Tune in to what is happening in those technologies that can help sustain an edge in a competitive conflict. Several choices exist, e.g., buying a technology, investing in startups, or partnering with a compatible company.

Such was the case with Comcast when it agreed to buy Time Warner Cable in 2014. With an evolving disruption expected in TV, the organization aimed to leave behind the safe confines of the cable business and become what is called a global digital platform. That means being able to function as an online video distributor capable of delivering a diverse menu of live and on-demand programming choices across multiple devices, such as TVs, tablets, and mobile phones, to consumers around the world.

Establish Strong Internal Communications

Maintaining ongoing and meaningful communications is founded on the rock-solid principle that outstanding leadership and managerial competence produce trained, motivated, and well-informed employees—the essential elements in preparing for and implementing a turnaround—and which was lacking in the J.C. Penney case.

To those ends, establishing internal channels of communications is built on the following five realities:

1. Implementing even the most basic competitive strategies requires well-honed leadership to activate the corporate vision and motivate employees for a turnaround.
2. Rigorous business principles do exist and need to be followed if a plan is to succeed, regardless of the type of business.
3. Effective decision making depends on nurturing a customer-driven company culture consisting of core values that must be communicated consistently to employees, if managers are to expect winning performance.
4. Business excellence relies on persistently searching for best practices, then communicating them selectively within the firm.
5. Looking for fresh opportunities requires managers to cultivate in their employees a wide-vision lens. As geographic distances dissolve, cultural differences emerge as the more important possibilities to integrate in a turnaround plan.

Thus, establishing ongoing communications provides you and others with the ability to inspire entire groups using various formats, such as motivational talks, news of corporate and individual achievements, and information about new products and organizational events.

Effective communications provide consistency and positive reinforcement that can influence employee behavior, strengthen a team culture, revitalize morale, and cultivate an entrepreneurial mindset.

Then, there is the pragmatic issue of updating employees' skills. These include timely messages to encourage employee self-development and equip them with the know-how to shape innovative strategies suitable for a turnaround.

STRONG VERSUS WEAK CULTURES

Within the organizational framework, strong cultures exist where the staff is aligned to the organization's values. In such environments, strong cultures help firms perform efficiently. Conversely, weak cultures have little alignment with organizational values, and control must be exercised through extensive procedures and supervision that often result in building layers of bureaucracy.

The benefits of a strong organizational culture include the following:

- Improved alignment of the firm with its strategic direction, objectives, and the rest of the firm's business plan
- High employee motivation
- Increased team cohesiveness among various business units
- Consistency in the form of command and control within the organization
- Improved corporate performance
- Increased proficiency in implementing offensive and defensive forms of confrontations

Numerous scholars have done extensive research on organizational culture. The more noteworthy concepts are digested in Table 2.2.

THE POWER OF MORALE

Morale acts as the second major force that forms the foundation of the psychological dimension of conflict. It supports unity within the group and taps the inner strengths of individuals, which aids members of the staff from wavering under the stresses of competitive confrontations. It is also the force to energize those with the resolve to act decisively at the critical moments of a turnaround campaign.

Morale, then, is significantly shaped by the prevailing values that make up the organization's culture. Where morale is at a high level, it results in a cohesive team effort, which reflects in a sheer determination to win.

Therefore, it is in your best interest to let your staff know that support exists at the highest levels of the organization. Also, they must see and feel

TABLE 2.2

Noteworthy Concepts on Organizational Culture

Organizational culture represents the collective values, beliefs, and principles of organizational members. It includes such factors as history, product, market, technology, and strategy, type of employees, management style, and national cultures. It also refers to those cultures deliberately created by management to achieve specific strategic ends (Needle, 2004).

Organizational cultures can be distinguished by values that are reinforced within organizations according to seven categories: innovation, stability, respect for people, outcome orientations, attention to detail, team orientation, and aggressiveness (O'Reilly, Chatman, and Caldwell, 1991).

Organizational culture is defined as the way things get done around here, and is illustrated by four different types of organizations:

- *Work-hard, play-hard culture:* This has rapid feedback/reward and low risk. Stress comes from quantity of work rather than uncertainty.
- *Tough-guy macho culture:* This has rapid feedback/reward and high risk. Stress comes from high risk and potential loss/gain of reward. Focus is on the present rather than longer-term future.
- *Process culture:* This has slow feedback/reward and low risk characterized as plodding work, comfort, and security. Stress comes from internal politics and absurdities.
- *Bet-the-company culture:* This has slow feedback/reward and high risk, resulting in high stress and delay before knowing if actions have paid off. The long-term view is taken with extensive planning (Deal and Kennedy, 1982).

Organizational culture is a set of shared mental assumptions that define appropriate behavior for various situations. In larger organizations, there are diverse and sometimes conflicting cultures that co-exist due to different characteristics of the respective management teams (Ravsi and Schultz, 2006).

Organizational culture is the most difficult attribute to change, outlasting organizational products, founders, and leadership and all other physical attributes of the organization. Three cognitive levels describe organizational culture:

- *Artifacts* include facilities, offices, furnishings, visible awards and recognition, company slogans, and mission statements. They also take in how members dress and how they visibly interact with each other and outsiders.
- *Rituals* are the deep-rooted interpersonal values and behaviors that constitute the fabric of an organization's culture. They include the language, stories, myths, as well as the basic beliefs, assumptions, and impressions about the trustworthiness and supportiveness of an organization.
- *Unseen elements* are those that are unspoken elements of organizational culture. They often exist without the conscious knowledge of the members. Concealed at the deepest level of the culture, these elements cannot be easily identified by interviews and surveys, and thereby are missed by those attempting to understand the organizational culture (Schein, 1992).

(Continued)

TABLE 2.2 (CONTINUED)
Noteworthy Concepts on Organizational Culture

Organizational culture can be described as a cultural web, identifying a number of elements:
- *The paradigm:* The organization's mission and values.
- *Control system:* The processes that monitor what is going on in the organization.
- *Organizational structure:* Reporting lines, hierarchies, and the way that work flows through the business.
- *Power structure:* Who makes the decisions, how widely spread is power, and on what is power based.
- *Symbols:* Organizational logos and designs, as well as those symbols that extend power, such as parking spaces and other privileges.
- *Rituals and routines:* Management meetings, board reports, and any activity that tends to become habitual.
- *Stories and myths:* Messages about people and events that convey the history and values of an organization (Johnson, 1988).

Organizational cultural change can occur through the following guidelines:
- Formulate a clear strategic vision of the firm's new strategy, shared values, and behaviors.
- Display top-management commitment to change by showing that it is being managed from the top of the organization.
- Model culture change through behavior that symbolizes the kinds of values and behaviors that should be realized in the rest of the company.
- Modify the organization to support organizational change by identifying what current systems, policies, procedures, benefits, and rules need to be changed to align with the new values and desired culture.
- Encourage employee motivation and loyalty to the company. Managers should be able to articulate the connections between the desired behavior and how it will impact and improve the company's success. Training should be provided to all employees to understand the new processes, expectations, and systems.

Develop ethical and legal sensitivity. This is particularly relevant as it relates to affecting employee integrity, types of controls, equitable treatment, and job security (Cummings and Worley, 2004).

your physical presence as often as possible. They need to absorb the psychological comfort and confidence that would sustain their morale and motivate them to keep trying, even under adverse conditions.

The remote or reclusive leader works at a distinct disadvantage in any competitive conflict. One senior manager refers to the process as "manage by walking around." Another intoned: "Don't expect what you don't expect."

Then, there is the iconic former General Electric CEO Jack Welch who was known to walk the factory floor enticing workers with his famous "call me Jack" invitation to talk. Those individuals often received personalized, hand-written notes after Welch's visit. Many were treasured for many years by the recipients. Yet, few if any of these valuable lessons seem to have been internalized by many senior executives during the crucial period of Penney's turnaround.

Specifically, as for unity of effort within the group, it is the collective efforts of individuals within teams interacting among themselves that cultivate unity, rather than any isolated actions. For that reason, it is your essential role to encourage such interaction (see Table 2.1).

In the end, it is up to you to know your people and understand the reasons behind such highly charged displays of behavior as anguish and fear or courage and dogged determination. Therefore, you cannot be a stranger to your employees.

It is your influence and physical presence that affects morale. If they feel themselves no longer supported, it creates an untenable situation for you. Thus, unity requires that you nourish your confidence and your ability to lead.

This is all part of developing a positive culture to guide your organization through the turnaround. Negative cultures, on the other hand, have made the headlines in past years, from the likes of Enron, by organizations within the financial community that were identified with the 2008 global financial collapse.

Consequently, your aim is to shape your company's or business unit's culture to support the organization's vision and create harmonious human relationships. Whereas today's executives seem to embrace the popular, *participative*, and touchy-feely style of leadership, there is a diametrically opposite approach with an *autocratic* style that is less prevalent among major organizations, but still finds a home in some top firms and with the individual styles of particular executives. (Ahead are summaries about the major contributors to behavioral business psychology.)

Another dimension associated with harmonious human relationships and morale is in reaching the hearts of individuals. Heart collectively describes the emotional qualities by which you lead. These include unity, camaraderie, purpose, duty, and hope. In day-to-day organizational life, emotions materialize in conflicting forms, such as through order or confusion, commitment or indifference, boldness or fear, loyalty or deceitfulness. These, too, are the realities that typically dominate the heart.

Not only does heart underlie your role as a manager, it reflects in your outward behavior and ability to perform as an inspiring leader.

Accordingly, a visible display of confidence, discipline, and purposeful direction filter down and impact your performance, and, subsequently, impact the attitudes of staff members who might be low on morale and stripped of courage.

On the other hand, should you lose nerve, the winning spirit, and decisiveness to complete the planned efforts, then your ability to lead under stressful situations is at serious risk. Heart, therefore, contains the highly humanistic components that profoundly impact your capacity to manage your people, your competitors, and your ability to implement a turnaround plan.

In the last analysis, success in a turnaround, or any encounter, is a matter of morale and reaching the hearts of your people. In all matters that pertain to an organization, it is the human heart that reigns supreme at the moment of crisis. Some careless managers rarely take it into account. And errors, sometimes irreversible, are the unfortunate result.

Table 2.3 summarizes conditions that can negatively impact morale.

What, then, are the principal psychological concepts behind the bursts of creativity and energy that can make the behavioral difference between success and failure? Even decades after their works were first published, the most distinguished contributors to behavioral business psychology remain high, such as Abraham Maslow's Hierarchy of Needs; Frederick

TABLE 2.3

Morale and Its Impact on Business Strategy

- Morale fades when there is no inspiring goal for which to fight.
- Morale weakens when employees work in a state of uncertainty and a negative mindset.
- Poor communications affect morale with a corresponding serious effect on implementing turnaround strategies.
- Lack of staff commitment interrupts the turnaround plan.
- Unethical corporate behavior affects morale and overall performance.
- An inability to create a long-term vision and develop attainable objectives.
- Undisciplined behavior at various levels that readily gives in to fear.
- A leader who procrastinates and visibly displays an inability to handle responsibility.
- A leader who stifles employees' input.
- A leader's failure to accurately assess the competitive situation.
- A leader who does not know how to reach the heart, mind, and spirit of a group, instead shows more interest in his personal agenda.
- Morale suffers and defeats become competitive disasters without sufficient training of leaders and staff.

Herzberg's Motivation-Hygiene Theory; Douglas McGregor's Theory X and Theory Y; and William Ouchi's Theory Z. Others include Victor Vroom's Expectancy Theory and Edwin Locke's Goal Theory.

A brief review of the essential ideas supporting each concept follows.

Abraham Maslow

According to Maslow's Hierarchy of Needs, there are five levels of needs that motivate people, which are usually displayed as a pyramid:

> First are the basic *physiological* needs that cover comfort and maintenance, such as food, drink, shelter, and health.
> Second are *safety* needs, which refer not just to physical safety and protection from harm, but also to such areas as financial security, legal assistance, and all the means that maintain stability.
> Third are *belonging* needs that indicate the need for human contact, such as with family, friends, relationships, and teams.
> Fourth are *esteem* needs, which recognize the need for status, power, prestige, acknowledgment, respect, and responsibility. Such wants motivate individuals to reach for a higher position within a group.
> Fifth are *self-actualization* needs, which indicate a desire to reach for his or her full potential and strive for individual destiny, after all previous needs have been satisfied.

Frederick Herzberg

Known as the Motivation-Hygiene Theory, the primary element is that the motivating factors are embedded in the satisfaction gained from the job itself. He reasoned that to motivate an individual, a job must be challenging, with sufficient scope for enrichment and interest.

Motivators—often called satisfiers—are directly concerned with the satisfaction gained from the job. Conversely, a lack of motivators leads to over concentration on what Herzberg called hygiene factors—or dissatisfiers—that form the basis for complaints. For instance, satisfiers include sense of achievement, recognition, opportunities for advancement, and status. Dissatisfiers involve quality of management relationship with manager, working conditions, wages, and interpersonal relationships.

Douglas McGregor

McGregor's well-known Theory X and Theory Y remain central to organizational development and organizational culture. He maintained that there are two fundamental approaches to managing people. Theory X tends to use an authoritarian leadership style. In contrast, Theory Y leans toward a participative approach.

Theory X is characterized as the average person's dislike for work, forced to work toward organizational objectives, prefers to be directed, and wants security above all else. Whereas, with Theory Y, an individual's work is natural and enjoyable, is self-directed to achieve organizational objectives, and often seeks responsibility.

McGregor places money in his Theory X category and feels it is a poor motivator. On the other hand, praise and recognition are placed in the Theory Y category and are considered stronger motivators than money.

William Ouchi

Ouchi's Theory Z is essentially a combination of all that is best about McGregor's Theory Y and modern Japanese management. It places a great amount of freedom and trust in the positive relationship between employees and leaders. It also assumes that employees possess strong loyalty and interest in teamwork and in the organization.

Ultimately, Theory Z promotes common structure and commitment to the organization as well as constant improvement of work. It also places a great deal of reliance on the attitudes and responsibilities of workers, whereas McGregor's XY theory is mainly focused on management and motivation from the manager's and organization's perspective.

Victor Vroom

Vroom's Expectancy Theory of motivation is based on the expectation of desired outcomes. "In general, people will work hard when they think that it is likely to lead to desired organizational rewards," according to Vroom. Specifically, the theory is based on three concepts: valence, expectancy, and force. Valence is the attractiveness of potential rewards, outcomes, or incentives. Expectancy is a person's belief that he/she will or will not be able to reach the desired outcome. Force is a person's motivation to perform.

Edwin Locke

Locke's Goal Theory describes setting specific goals as a means to obtain higher performance. He believed that, through employee participation in goal setting, the employees would be more likely to accept the goals and have greater job satisfaction. The underlying assumption is that employees who participate in goal setting will set more difficult goals for themselves and yield superior performance.

Other motivational theories exist, with each taking somewhat varying approaches. Virtually all agree, however, that a motivated worker seeks better ways to do the job, is generally more quality-oriented, and overall is more creative and productive. And where leadership is slipshod with little visible motivation and little effort given to strengthening relationships, morale declines, quality of work deteriorates, and competitive confrontations take on greater risk.

RELATIONSHIPS BETWEEN LEADER AND STAFF: EXPECTATIONS FOR DEVELOPING A TURNAROUND PLAN

How do the above discussions on organizational culture, morale, and motivation translate into meaningful turnaround strategies? The following sections present ideal expectations from staff in planning a turnaround.

Expect Active Participation from Staff

As an extension of the earlier comment about seeking input from all levels of your employees, are you able to maintain an ongoing interchange with your staff, so that they become familiar with the overall strategic direction of your thinking? That means entering into a dialog so that your staff understands the nature of the turnaround plans. In particular, such awareness on both sides is significant should you have to wait before implementing your turnaround plans due to sudden market events.

In that instance, you are likely to find an exaggerated level of anxiety, not only within yourself, but with your staff. This is where you, and they, must cool down and learn to wait for better timing, even though events are still in the midst of a crisis. This is the moment, despite all determined efforts, to exercise disciplined patience.

The essential point: Between acceptance and anxiety, choose acceptance, with the proviso that patience is only a pause, nothing more, that allows you time to reignite your efforts with fresh ideas and innovative strategies to push forward with your plan.

Therefore, help your people create compatible relationships with others. In some instances, you may have to take on the role of an active mediator. Such compatibility isn't always completed. Certain managers will back off and ask the opposing groups to work through their problems. Notwithstanding, it is your obligation that once anger, frustration, and continuing dispute move forward beyond a reasonable time, you have to take an active role in setting a pathway for conflict resolution.

Still other executives intentionally create opposition in their ranks, thinking that it leads to a healthy work environment. However, it is with a degree of risk, when compared to the far greater benefit of creating harmony and unity of effort. Nonetheless, strained relationships do exist among various functional areas of an organization, such as marketing versus finance, product development versus distribution, manufacturing versus sales, and so on.

For the most part, however, hostilities are counterproductive, especially when flaring anger dominates the scene. Clashing groups seldom arrive at an acceptable solution. Within such a fractured situation, your aim is to promote workable outcomes based on internal operating conditions, the dynamics of the marketplace, and the activities of competitors that may be beating hard against you.

As part of a healing effort, utilize a workable system of rewards and recognition. Such a system includes much of what is generally accepted about positive reinforcement and classic motivational behavior. Look again at those concepts identified with Herzberg, McGregor, Maslow, Ouchi, and others.

Further, as part of the process of nurturing your staff to heightened levels of performance, initiate (or recommend) procedures to capture the insights, knowledge, and observations of numerous individuals and categorize them into usable databases. Such information then should be available and easily accessed for tapping the full range of experiences and knowledge of your staff.

Expect Staff to Maintain Momentum

It is important to maintain momentum and keep any possibility of a successful turnaround on your side. Don't think of your actions, or those suggested by your staff, as unreasonable or impossible. The essential point is that it is always reasonable to expect that ideas, alternatives, and possibilities do exist.

This means leaning heavily on the quality of your personnel. As already emphasized, this is possible where you address key issues affecting their morale, such as activating campaigns that lead to regaining growth and creating a feeling of unity within the group. These areas have as their underpinnings spirited leadership on your part, a clearly understood vision or objective, and astute competitive strategies.

To turn a potential failure into a successful turnaround depends greatly on continuing training and making certain all the attributes of a healthy culture cited above come into play. Blended together, they give confidence to individuals to carry on in spite of any inclination to concede defeat.

Where are the behavioral aspects of conflict most visible? Consider such turnaround campaigns to reclaim a former market position, defend a market position that represents a major source of revenue, expand into new markets, or confront the aggressive moves of a competitor trying to weaken your resistance.*

Within these campaigns, there is the psychological dimension that deals with the subtleties of creating conditions whereby the slightest prospect of defeat might be enough to cause one side to yield. And, where there is a mutual interest on both sides to create a balance that looks to avoid further confrontation, then a live-and-let-live attitude might prevail, thereby giving you breathing space to build back your offensive capabilities.

Such a situation may come about in a rather understated fashion. If one firm senses that the rival doesn't really want to create a serious confrontation, which is likely to last for a lengthy period and consume excess resources, then the meeting half-way approach will prevail.

How, then, can these conditions come about? One workable approach is to choose objectives that will not intimidate the rival, or attempt to dislodge him from the marketplace, or use aggressive actions, such as drastically reducing prices and upsetting the equilibrium of the market. It also means not launching overly aggressive promotions with the clear aim of taking away customers or penetrating the rival's primary segments.

Thus, the psychological aim is to create a condition where the competitor's personnel are disheartened, lack enthusiasm, and can no longer carry on the conflict. Even using pretense to create an inner fear among individuals is enough to cause a sharp reduction in the rival's effectiveness. In this manner, even an inferior force can exist in the midst of formidable adversaries. All this activity is part of reducing the competitor's effectiveness and thereby neutralizing his capabilities.

---

* It should be noted from the J.C. Penney case that virtually all of these types of campaigns were not executed successfully.

Expect Staff to Neutralize Competitor's Strategies

One of the major turnaround strategy decisions is how to make the conflict more costly and burdensome for the rival. The central question, then, is why use up one's own resources if victory can be attained without draining your organization?

That means creating conditions that would result in the competitor's expending (or wasting) its human, financial, and material resources in unplanned moves, with little opportunity to replenish those resources. In effect, placing a rival in this condition is another way of neutralizing it without an actual confrontation.

An example of this wearing-down process is shown in the case of Samsung Electronics (also referenced in Chapter 1). In the area of kitchen appliances, Samsung's laudable objective is to dominate the market and capture no. 1 market position. That means going up against such established brands as Whirlpool, General Electric, and Kenmore, brand names known to consumers for decades.

With that goal is the broader strategic objective of becoming the world's largest appliance manufacturer. By 2014, the Korean company had already become the fastest growing appliance brand in the United States. Using its experience and expertise in technology as its dominant strategy, Samsung developed refrigerators, washers, and LED lighting that allows users to remotely operate them through smartphones and tablets. The appliances also connect to the Internet independently to download new software or let users monitor their homes via built-in exterior cameras.

Relying on its smartphone designs, Samsung pushed appliance-specific features into each product to differentiate them from the competition. Its *Chef Collection* refrigerator, for instance, can dispense still and sparkling water, keep different zones cooled to different temperatures, and convert a small fridge compartment into a secondary freezer. Thus, the applications go deep into the core of Samsung's technology and branding strategy as it utilizes its unique capabilities in phones, TVs, and semiconductors in its quest to build the smart home.

For Samsung, there is a tangible effect of wearing down the competitors by means of superior technological, financial, and psychological capabilities. The longer the duration of this condition, the more assured will be its success.

As for competitors, Samsung's powerful thrust into the normally slow-moving appliance industry had to send shock waves through the corridors of all those organizations. For the smaller ones, just attempting to catch up would take immense resources to compete. And, the more delayed the response, the harder will be the effort. For the larger contenders, budgets may have to be recast to make the appropriate investments.

In all cases, however, those companies hard hit by the assault from Samsung may have to face a turnaround by either leaving the market or adopting a niche strategy using specialized applications, technical innovations, price points, product refinements, or logistical advancements as a means of survival.

In still other situations, the remaining firms would have to consider joint ventures with organizations that can bring the best technology to the basic appliances. In each instance, these are displays of resistance. Depending on how each is carried out, there is always the possibility of growing and finding a new decisive point for a turnaround.

Otherwise, attempting a pure self-defense means fighting with the goal, as noted above, of buying time until an opportunity occurs. Often, that policy can be pushed only so far and then it turns into wishful thinking as the company's presence in the market continues to shrink.

Yet, even under those circumstances, there is the chance of the competitor faltering, its products nonperforming, service mishandled, and the like. The aim, then, is to be as vigilant as possible to see areas of weaknesses and seize the moment to exploit.

The significant point: There are numerous options in planning a turnaround, from outright confrontations to passive resistance as shown in the following guidelines:

- Pursue revenue-expansion opportunities as well as cost-reduction opportunities. By latching on to new systems and disruptive technologies, it is possible to create a meaningful competitive advantage, and possibly neutralize the opponent's capabilities.
- Position yourself in the market through rapid maneuvers, so that the competitor cannot anticipate your moves in sufficient time to counter your actions with a significant defense.
- Focus your resources on an emerging, neglected, or poorly served market. Your aim is to avoid a direct, head-on confrontation with a stronger rival. An alternative aim is to cause the competitor to spread his resources by attempting to anticipate your moves, thereby weakening his primary efforts.

- Create a differentiated product, or value-added service, that is not easily cloned.
- Develop a system that provides accurate market intelligence, so that you can take fast action against market opportunities; as important, it allows you to react quickly to any areas of a competitor's vulnerabilities.
- Initiate constructive relationships with customers that lock out competitors for an extended sales cycle.

Expect Innovative Thinking

A leader just can't order up innovative ideas and creative solutions to a problem. However, you can influence behavior through words, actions, and by the type of working environment.

A case in point, Google creates a working climate in which its managers display the outstanding qualities of leadership by motivating their staffs to innovate in all aspects of their jobs. Recognizing that inventiveness and innovation are the drivers of organizational success, leadership is dedicated to creating a working culture that encourages fresh ideas.

For instance, Google management gives all engineers one day a week to develop their own pet projects, no matter how far from the company's central mission. If work deadlines get in the way of those free days for as much as a few weeks, they accumulate. Also, the system is so pervasive that anyone at Google can post thoughts about new technologies or businesses on an ideas mailing list, available company-wide for inspection and input.

What are the leadership traits that support such behavior? First, respect for the individual forms the basis of Google's leadership. In practice, it means recognizing and appreciating the inherent dignity and worth of people, and even where some individuals' ideas will not succeed, their efforts are recognized and respected. This is especially relevant working with culturally diverse personnel with a wide range of ethnic and religious backgrounds.

Second, at each level, leaders stand aside and let subordinates do their jobs. They empower their people, give them tasks, delegate the necessary authority, and let them do the work.

The fundamental issue here is that the organization is not going to stop functioning because one leader steps aside. Therefore, central to the job of good leaders means helping subordinates grow and succeed by teaching, coaching, and counseling.

Expect Staff to Stay Alert to Competitive and Market Conditions

Alertness means getting your staff to a point whereby they grasp how competitors' actions could challenge their ability to carry out a turnaround campaign. It also means being able to interpret market events with some accuracy and then adjust strategies to meet changing conditions, especially where it comes to determining from market intelligence where to find a decisive point of weakness in a competitor's offensive moves (see Best Buy case in Chapter 1).

Your intentions are to get your employees to understand the broader competitive world. Doing so, takes ongoing and carefully crafted communications. Increasingly, a wide variety of communications vehicles permit you to initiate programs and exchanges of information.

Such is the case with the burgeoning applications of social networks, blogs, email, instant messaging, video conferencing, corporate wikis, and the like. Also, face-to-face briefing sessions, thought to be outmoded by electronic communications, are still prevalent and noticeably on the increase.

Using any of the above forms of communications increases awareness and serves several specific purposes, among them:

- You keep your staff informed about meaningful market and competitor events.
- You motivate them by providing a venue to participate with fresh insights.
- You tap their diversity, thought patterns, experiences, and collective knowledge, thereby integrating individuals and their functions.
- You obtain viewpoints that can provide useful perspectives and constructive comparisons.
- You bring unity to your group.

All of this interaction helps you to stay close to your employees with the intention of sensitizing yourself to their thoughts, moods, temperaments, and increasingly the nuances of their thoughts, and then integrate them into the mainstream of your company's goals. Your overall aim is to focus on sustaining their psychological well-being. As a result, you should see tangible improvements in performance, innovation, and employee harmony.

Expect Staff to Respond to Negative Behavior

Within the context of psychological well-being, substantial harm to employee morale can result if negative forms of behaviors are left unattended. It is in your best interest to make every effort to break those mannerisms, which can show up as hostile feelings, fears, and a loss of confidence. That means, in your managerial role as coach and counselor, do all you can to convince them not to make the same mistake as others who prematurely gave up trying to change because they felt stuck in their behavior.

Demonstrate to them that if they truly want to they can overcome undesirable forms of behavior. Here is where discipline plays a positive role. In effect, it is a test designed for them to exercise freedom of choice. The issue, then, is for them to choose wisely.

Your best approach is to set up (or recommend) a program to deal with negativity. Where possible, utilize outside specialists. Specify a clear-cut objective to help individuals resist pessimistic thoughts. Often, they can direct the process to clear the blockages that prevent implementing your turnaround plan.

By taking such positive action to shape positive relationships, you acknowledge that your people are a major influence in market performance. And, as important, they function as key competitive differentiators.

Finally, as you prepare your organization or group for the turnaround, avoid the unnecessary search for shortcomings and weak points in yourself and employees, unless there is a deliberate effort to conduct a needs analysis for future training. Instead, focus on their good attributes. Highlight them and turn even indifferent employees into winners. Keep in mind, too, they are the ones who will carry out your plan, which has at stake your personal reputation—and perhaps your job.

In the process, be alert for the inevitable expressions of hardship, complaints of tough work, and sacrifices that often surface among workers during any emergency. Distinguish between what you would consider grueling work and merely a display of whining and grumbling. To counter any emotionally charged outbursts, indicate to the rest of the staff the consequences if they continue with their flare-ups.

If you succeed in choking off their complaints, and, as a result of their sacrifices and hard work, you show first-rate results, point out that those positive results are only the beginning of other good outcomes.

It is your opportunity to say, "If you call this good, I'll really show you what good is. I'll show you how this company can grow and how each of you can benefit financially and professionally." Of course, this approach must be anchored to tangible and realistic assessments that the staff can understand, believe in, and rally around.

What ties into that approach are the opportunities that emerge each day. Some are quite apparent, others require more searching. Also, what might appear as sameness could be an opportunity under a different set of market conditions.

As long as there is a mindset tuned to fresh opportunities, it is your job to instill in your people the ability to seek possibilities out of what every day and each event has to offer. You must be sure, however, that your employees see the opportunities, agree with them, and be of the same opinion.

Workers will be happy to change their individual behavior if they understand why and how their actions contribute to the overall company's fortunes. Also, they will act positively if they believe it is personally worthwhile for them to play an active role in the organization. Even where you cannot affect change in the organization, it is possible to take the initiative and set up a system within a small group.

Now, move on to Chapter 3 (Prepare a Turnaround Strategy Plan).

REFERENCES

Cummings, Thomas G. and Worley, Christopher G. (2004). *Organization Development and Change*, 8th Ed., South-Western College Pub.

Deal, T.E. and Kennedy, A.A. (1982, 2000). *Corporate Cultures: The Rites and Rituals of Corporate Life*, Harmondsworth, Penguin Books, 1982; reissue Perseus Books, 2000.

Johnson, Gerry (1988). "Rethinking Incrementalism," *Strategic Management Journal*, Vol. 9, pp. 75–91.

Needle, David (2004). *Business in Context: An Introduction to Business and Its Environment*. ISBN 678-1861529923.

O'Reilly, Chatman & Caldwell (1991). "People and organizational culture: A profile comparison approach to assessing person-organization fit." *Academy of Management Journal*, 34, pp. 487–516.

Ravasi, D. and Schultz, M. (2006). Responding to organizational identity threats: Exploring the role of organization culture. *Academy of Management Journal* 49(3), 433–458.

Schein, Edgar (1992). *Organizational Culture and Leadership: A Dynamic View*. San Francisco, CA: Jossey-Bass, p. 9.

3

Prepare a Turnaround Strategy Plan

Chapter Objectives

Be able to

- describe the components of a turnaround plan;
- develop a statement of strategic direction;
- select objectives;
- identify turnaround strategies; and
- prepare postcampaign strategies to secure a turnaround.

INTRODUCTION

Chapters 1 and 2 established five essential concepts underlying the successful development of a turnaround plan.

First, determine the primary conditions that triggered a need for a turnaround plan. For instance, an aggressive company attacks a defender that is unable to protect its vulnerable position. This point was illustrated by Panasonic when it spread its resources over too large a market area while attempting to maintain an overly broad number of product lines. The organization was unable to adequately mount effective defenses.

Then, there was the instance of Kodak falling into a state of complacency, relying on old technology, and refusing to recognize in time the industry's move to digital photography. (Recent reports indicated that the one-time industry leader relied heavily on a false sense of security; that its reputation and long years in the market would create adequate barriers.)

Next was the case of Intel recognizing an impending crisis and rushing to take rapid action. It reorganized the firm and immediately focused resources on a single paramount objective: Latch on to a prevailing industry trend by reorienting its product line to focus on tablets, smartphones, and similar devices to regain a leadership position in microchips.

The second concept, underscored by Intel, conveyed the idea that the external market condition that resulted in an internal crisis didn't occur from some random incident. Consequently, what follows the problem/identification phase is initiating remedial actions to identify and root out endemic obstructions that created the predicament.

For instance, some areas in which to immediately look for deficiencies include lack of a competent strategic business plan, ill-trained staff, ineffective market intelligence, demoralized personnel, ineffectual leadership, a flawed organizational culture, or a cumbersome decision-making process. A more expansive listing of factors is provided at the end of this chapter (Postcampaign Strategies to Secure a Turnaround).

Thus, relying on the often-used approach of beginning the turnaround by indiscriminate cost cutting, eliminating product lines, and shelving new projects doesn't always address the underlying causes of a crisis. Although pragmatic attention should be given to each cost item and every project, an essential step in the turnaround process is uncovering preexisting conditions, ones that may be pervasive, hidden, and long-standing within an organization, as in the case of General Motors (GM is also cited in Chapter 1).

A third concept focused on developing strategies to neutralize the competitor's capabilities from doing you further harm. That means recognizing the various types of campaigns, any of which you are likely to face.*

The fourth concept highlighted the need to understand the vagaries of chance and luck that can bring a turnaround effort to a standstill. These intangibles often create psychological problems—notably, anxiety among management and staff.

In turn, such emotional reactions can enter individuals' minds and form negative attitudes that often bring about a paralysis of the mind that affects performance. Thus, there is the ever-present cloud of possibilities and probabilities that hover around the planning process where an atmosphere of undesirable risk and gamble exists.

The fifth concept emphasized developing a well thought-out turnaround strategic plan, which consists of *strategic direction, objectives, turnaround*

---

* See Chapter 1 for a list of 13 types of campaigns, as well as Table 1.1 for a source of approaches to neutralize the competitor.

TABLE 3.1

Turnaround Business Planning Process

A. Preplanning Activities

1. Identifying primary conditions triggering a turnaround	2. Uncover preexisting problems within the organization	3. Begin remedial action	

B. Strategic Planning Steps

4. Strategic direction	5. Objectives	6. Turnaround strategies	7. Postcampaign strategies

strategies, and *postcampaign strategies.* This format, then, provides the underpinnings of your turnaround. A flow chart of the process is shown in Table 3.1.

Once corrective action has begun and emergency actions have been taken to stop further losses, the turnaround plan can begin taking shape. The first step is to develop a strategic direction. This is a key move in that it reflects the policies of the organization and provides a description of the firm's long-term outlook.

Parts of the input are the demographic and psychographic (behavioral) analyses of various market segments. Armed with that data, objectives and strategies are developed that give life to the plan.

As for the tangible outputs of the plan, these reflect operations within *two zones of activity*: the first zone consists of product and service offerings that fulfill what the late management scholar Peter Drucker describes as the object of business: "to create a customer;" the second zone indicates strategies to neutralize the competitor's ability to do you harm or prevent you from achieving your planned objectives.

ESTABLISHING A STRATEGIC DIRECTION

With a strategic direction, you look at the big picture, which most often has a timeframe of at least five years. It is one of the foundation pillars of a turnaround plan. If thoughtfully developed, the strategic direction permits you to focus with a perceptive inner eye that takes in the entire market and competitive scene (noted above as two zones of activity).

56 • *Developing a Turnaround Business Plan*

The process expands your intuitive insight into what your organization or business unit would look like over the long term; provides the direction that your strategies are likely to take when engaging rivals in a conflict, determines how far you are willing to risk your company and its resources, and aids in selecting the type of offensive or defensive efforts you are likely to undertake, and, in the context of this book, gives you a direction to turn around and pull out of the crisis.

The following questions provide an organized approach for you to develop a strategic direction. Your precise answers will help shape a clear vision of what your company, business unit, or product/service will look like over the long-term planning period.

Implicit in these questions are issues related to senior management's view on how big to grow; how fast to expand; what people skills are required; and, of course, the availability of financial, material, technology, and other resources.

Also, within the framework of this book on developing a turnaround, is how to react to competitive confrontations where defending a market position is at stake, or where expanding into a competitor's territory is the objective.

Let's examine each of the questions in detail:[*]

1. *What are our organization's distinctive areas of expertise?*
   This question takes some soul searching to track the DNA that makes your firm different and unique. It also will take some pragmatic examination to see how your firm stacks up against a variety of factors, from management strength and employee morale to technology and financial resources. (See Chapter 1, Table 1.1 for mind-jogging references.)
2. *What business(es) should we be in over the next five years?*
   This requires looking out beyond your existing business to examine, and even reimagine, what is on the horizon in new technologies, industry trends (see the Kodak case), competitive moves, and evolving market needs. It means deciding what your business should be on the familiar scale of: *Are you in the railroad business or*

---

[*] You can fine-tune the questions to fit the primary issues relevant to your company and industry. Also, where time and resources permit, use a cross-functional team to conduct an extensive SWOT (strength, weakness, opportunities, threats) analysis.

transportation business?* The implications are that if your business is too narrow in scope, the resulting product and market mix will be generally narrow and possibly too confining for growth; that is, too myopic and, thereby, restricted by an inability to grow.

On the other hand, defining your business too broadly can result in spreading capital, people, and other resources beyond the capabilities of the organization and your ability to defend it against aggressive competitors, as in the Panasonic example.

Therefore, as part of the process of defining what business you should be in, you have to take into account the culture, skills, and resources of your organization. It also means considering such factors as customer needs, business functions to be enhanced or added, and types of new technologies you need to compete successfully.

3. *What segments or categories of customers will we serve?*

   Customers exist at various levels in the supply chain and in different segments of the market. At the end of the chain are end-use consumers with whom you may or may not come in direct contact.

   Other customers within the distribution channel serve as intermediaries and typically perform several functions. Intermediaries include distributors who take possession of the products and often serve as a warehousing facility.

   Still other intermediaries repackage products and maintain inventory control systems to serve the next level of distribution. There also are value-added resellers that provide customer service, technical advice, computer software, or educational programs to differentiate their products from those of competitors.

   The essential point of this fundamental overview: Examining the existing and future needs at each level of distribution helps you project the types of customers you want to target for the five-year period covered by the strategic portion of your plan. It also pinpoints barriers or points of competitive opposition you have to overcome. These, too, may be the focal point of your turnaround strategy.

4. *What additional functions are we likely to perform for customers as we see the market evolve?*

   As competitive intensity increases worldwide, each intermediary customer along the supply chain is pressured to maintain an

---

* The reference is from Theodore Levitt's classic article, Marketing Myopia, *Harvard Business Review* Sept/Oct., 1975, p. 28.

advantage. This guideline question asks you to determine what functions or capabilities are needed to solve customers' problems.

More precisely, you are looking beyond your immediate customer and reaching out farther along the supply chain to identify those functions that would solve your customers' *customers'* problems. Such functions might include providing computerized inventory control, after-sales support, quality control programs, financial assistance, or rapid delivery.* Overall, this question relates again to the "myopia" issue that defines the business in terms of customers' needs, wants, and solutions to their problems.

5. *What new technologies will we require to satisfy future customer needs and to meet and exceed those used by competitors?*

Within the framework of the previous question and the practices of your industry, examine the impact of technologies on customer retention. Look at where your company ranks with the various technologies and types of software used for product design and productivity, manufacturing, and logistics.

Look, too, at the continuing changes in information technology and business intelligence with their disruptive effects on product innovation and market competitiveness. Also appraise such technologies as expert diagnostic systems for problem solving, the rapidly changing communications systems to manage and protect an increasingly wireless enterprise, and methods for dealing with the threats of damaging cyber attacks.

6. *What changes are taking place in markets, consumer behavior, competition, legal/environmental, and economic issues (global and local) that will impact our company?*

This form of external analysis permits you to sensitize yourself to those critical issues that relate to markets, the industry, as well as to existing and emerging competitors. This broad-based question is open-ended and can cover the changes that are most likely to affect your long-term view of the marketplace. They can be as narrow as local economic conditions and as broad as pending governmental regulations.

Specifically, for competition, the analysis considers the likelihood of confrontations among entire companies or their individual business units and product lines. Keep in mind, too, that campaigns can be

---

* In 2014, amazon.com announced bold plans for using drones to expedite delivery to its customers.

forceful acts to persuade a rival to relinquish control of a market segment or a territory.

Therefore, as part of the analysis, look at the managerial issues that affect employees' ability to react to competitive conflict, such as discipline, training, leadership, communications, and the underlying culture of the organization.

Deliberating on these six questions personally, and ideally with a cross-functional planning team, allows you to make a long-term visionary inquiry that becomes the underpinnings of your strategic direction. In turn, it guides your decision making when faced with competitive conflict.

The following is a statement of an actual strategic direction from a division of a large healthcare organization.

> Our strategic direction is to meet the needs of consumers and healthcare providers for drug-delivery devices by offering a full line of hypodermic products and product systems. Our leadership position will be maintained through internal research and development, licensing technology, and/or acquisition options to provide alternative administration and monitoring systems.

The division's primary product is the hypodermic needle. The strategic direction could have stated simply that it is a manufacturer of hypodermic needles. That would have been far too "myopic" and restrictive for growing and maintaining a dominant competitive position. It could have been the root cause of a crisis leading to a call for a turnaround.

The broader interpretations of "drug delivery devices" and "product systems" certainly incorporate its core business of hypodermic needles. The important issue, however, is that it excites the mind to create fresh opportunities for product designers to develop new products and services.

For example, systems and devices would include new forms of pills and internally implanted pumps with sensors to control the release of the drug within the body. Other devices that look like writing pens with drug-filled cartridges are alternatives to the syringe and needle. And still other product systems incorporate monitoring devices to measure medical effectiveness of the drug and automatically calculate the amount of dosages required for an individual patient.

Given an extra measure of creativity and a broader interpretation of the strategic direction could impact on markets, products, technologies, and services. For instance, safe disposal systems for needles and syringes linked to increasing environmental concerns; product configurations by types of diseases, geographic location, culture, and demographics within a target population segment. Then, there could be systems that calculate for the severity of an illness and its contagious impact on unprotected groups.

Therefore, by conceiving a broader interpretation of a business—from railroad to transportation—helps avoid the negative impact of a shrinking market position due to older technologies. Taking time to develop a well thought-out strategic direction provides an organized framework to extend your thinking to what your company or product line can or should become within an achievable time frame.

Look again at the strategic direction. In addition to product systems and services, the statement refers to a "leadership position." The division's existing position certainly could be maintained through internal R&D. However, the broader thinking also opens a pathway to new products, devices, systems, and services through licensing of technology, acquisition, joint ventures, and a variety of other forms of strategic alliances.

Although the example used here represents a division of a company, the same thought process is appropriate for a product line within a midsize organization, and certainly within a smaller single-product company. Again, the one danger is going too far without examining if adequate reserves are available to defend all market segments and products, which was the situation that affected the previously mentioned Panasonic.

OBJECTIVES

Once the strategic direction has been accepted, the next step is to develop objectives to give tangible form to the plan. There are two primary guidelines that should have your attention.

First, your objectives should indicate what outcomes you want to achieve and how they would relate to your strategic direction. Second, you should address the issues and preexisting conditions that triggered the need for a turnaround. The remedial actions should be inserted as well in the strategy section of the plan.

For clarity, divide your objectives quantitatively and nonquantitatively. Again, your timeframe is five years or whatever time period your organization requires.

Quantitative objectives indicate performance metrics, such as sales growth, market share, return on investment, profit, and any other quantitative objectives required by your management.

Nonquantitative objectives would span such diverse areas as initiating changes in organizational design, reexamining corporate culture, upgrading relationships within the supply chain, consolidating a segment position, establishing strategy teams, building specialty products to penetrate new markets, improving competitive intelligence systems, improving staff morale, upgrading technology, and any of the issues identified in Chapter 1, Table 1.1.

In the healthcare example, the division indicated its quantitative objectives related to sales, market share, and financial requirements. Then, there were the nonquantitative objectives, as noted by a sampling of the division's objectives:

- Maintain position as the low-cost producer while introducing new improvements to existing products.
- Aggressively maintain our dominant market-share position in all market segments.
- Maintain sufficient manufacturing capacity to absorb our competitors' market share in existing segments, as well as to serve new and emerging segments.
- Strengthen an intelligence-gathering network to serve as an early warning system of sudden changes in competitors' activities.
- Neutralize (competitor's name) ability to enter the home care segment of the market.

Note, too, how these objectives have long-term strategic implications. Where possible, you can add quantitative information for each objective. However, it is not always necessary in this strategic section.

Quantitative details can be added later in the plan, usually in the growth strategy section and certainly at the tactical one-year portion of the plan, where details are given, dates specified, and responsibilities assigned. The essential point is that this planning format permits flexibility to accommodate to the practices of entire organizations as well as to their individual business units and product lines.

Thus, the aim of the strategic direction is to provide a "vision" of what the future of the organization or group can look like; the objectives delineate the precise outcomes, strategies indicate the actions.

STRATEGIES

A major component of the plan, which covers significant portions of the following chapters, focuses on strategies as action elements that set the plan in motion. This section, then, outlines the means by which you achieve your objectives.

Whereas the emphasis of this book is at the strategic level with a timeframe of five years, you may find that an urgent need persists to take immediate action, particularly in a turnaround situation where timing is the pragmatic issue and artificial structures of timelines fade away in favor of survival.

Your thinking about strategies generally emphasizes the following:

- Identifying decisive points in which to concentrate resources against a market segment.
- Specifying actions to neutralize the competitor's capabilities.
- Selecting market segments that are best suited for beginning a turnaround, which open possibilities for regaining the offensive.
- Pinpointing areas that would represent competitive advantages for defending existing markets from aggressive rivals.
- Initiating remedial actions to deal with the root causes of the crisis.

At its most pragmatic level, think of strategies as actions to achieve your longer-term objectives, tactics as actions to achieve shorter-term objectives. In a broader sense, strategy is the art of coordinating the means (money, human resources, and materials) to achieve the ends (profit, customer satisfaction, and company growth) as defined by company policies and objectives.

Further, strategy consists of *actions* to achieve *objectives* at three distinct levels:

1. *Corporate strategy.* At this level, strategy is developed at the top echelons of the company or major business unit. The aim here is to deploy resources through a series of actions that would fulfill the vision and objectives as expressed in a long-term strategic business plan.
2. *Midlevel strategy.* At this juncture, strategy operates at the department or product line level. Its timeframe is more precise than corporate strategy. Typically, these strategies define actions for a long and

short time period, and they are more precise in describing specific objectives.
3. *Tactics.* This level requires a shorter time frame from those at the two higher levels. Normally, it links with a company's or product line's business plan, marketing plan, and annual budget. In everyday application, tactics are actions designed to achieve short-term objectives, while in support of longer-term objectives and strategies.

Tactics cover such areas as social media, sales force deployment, supply chain methods, customer relationship programs, training, product branding, value-added services, and the selection of market segments to launch a product or dislodge a competitor.*

Techniques and formats for indicating strategies can vary. Where some objectives tend to be longer term and broader-based, you may need to develop multiple strategies for each objective. For instance, you can use general strategy statements and then follow by restating the specific objective along with the related strategies.

For example, with the healthcare firm previously referred to, a format would look like the following:

GENERAL STRATEGY STATEMENT

The division will maintain industry leadership by addressing the full range of consumer and healthcare provider needs related to drug delivery devices. These include not only the marketing and delivery devices, which are virtually painless, easy to read, and convenient to use, but also programs and educational services to aid in the achievement of normal bodily functions to maintain overall good health. These additional services will meet user needs both at the time of diagnosis and in the continuing treatment of the problem.

A dominant position in the drug delivery device market will be maintained by developing market segmentation opportunities through continued product differentiation and innovation.

Objective 1: Maintain our low-cost producer status while introducing new improvements to existing products.

---

* Details about short-term tactics go beyond the scope of this book. Those topics are adequately handled by the specialized books and articles on those varied subjects.

Strategies:

- Reduce costs by 32.5 percent before 202X. Maintaining low-cost producer status gives our company the widest strategic flexibility in dealing with competitive assaults on our franchise.
  Potential areas of cost reduction:
    - Overhead reductions, 4.5 percent
    - Waste reductions, 7.0 percent
    - High-speed needle line, 6.5 percent
    - Sales territory redesign, 8.0 percent
    - Quality improvement and reduction in repair service, 4.0 percent
    - Packaging improvement, 2.5 percent
    Total: 32.5 percent
- Upgrade existing products through improved dosage control and enhanced packaging to maintain a competitive advantage.

Objective 2: Aggressively maintain our dominant market share position in all market segments.

Strategies:

- Develop the next generation of Supra-Fine needles to maintain superior product quality and performance versus competition, as it relates to injection comfort.
- Increase spending levels on consumer/trade support programs to provide added value to product offerings, thereby decreasing attractiveness of lower-priced alternatives while maintaining brand loyalty.
- Maintain broadest retail distribution and highest service levels to gain retailer support in promoting our brands and carrying adequate inventory levels.
- Continue healthcare educational programs to gain professional recommendations at time of diagnosis and thereby maintain brand loyalty among users.

Objective 3: Neutralize (competitor's name) ability to enter the home care segment of the market.

Strategies:

- Preempt the competitor's introduction of its new system with our new multiple dose therapy.
- Convert users to a 40-unit syringe and cancel the competitor's advantage and neutralize its campaign.

- Develop and introduce a disposable pen-cartridge injection system to further segment the market and thereby reduce the competitor's advantage as a point of entry.
- Become a full-line supplier of drug delivery devices by broadening product offerings through internal research and development, joint venture, licensing, and acquisitions, thereby blocking several entry areas.

As noted in the above, the strategies of that healthcare division cover a wide range of activities and incorporate a variety of functions within the organization.

Accordingly, there is a practical necessity for involving as many functional managers as possible in developing your plan. Not only will their ideas prove helpful, they also will internalize the strategies and be more motivated to implement them.

Implementation is thereby achieved through managers from manufacturing, product development, finance, sales, and distribution participating in making the turnaround plan come alive. This type of collaboration is demonstrated with highly successful results with such organizations as Toyota, Whole Foods, and IBM.

POSTCAMPAIGN STRATEGIES TO SECURE A TURNAROUND

This final section of a turnaround plan involves a high level of vigilance. The essential point is to assure that the circumstances causing the original crisis are not repeated.

Expressed yet another way, complete attention is required to interpret signs of internal and external conditions that could turn into another emergency. That requires astute leadership to oversee a diverse range of issues.

For instance, that means identifying the various types of competitive conditions that would block you from securing your turnaround. Another is watching for signs of friction that could prevent your campaign from moving forward. Then, there are the ever-present ambiguities of chance and luck that seem to appear unexpectedly. All these unknowns suggest remaining highly flexible to deal with the "fog of war,"—the natural elements of competitive conflicts.

Another major task is for you to be fully aware that a poststrategy requires top-to-bottom planning that integrates the thinking and contributions of various members of the staff, as was illustrated in the healthcare case. Doing so helps align the organization's culture with the turnaround plan and, thereby, contributes to elevating morale through a unified staff. (These issues are covered in detail in Chapters 1 and 2.)

What, specifically, are the danger signs that could prevent you from securing a turnaround with poststrategies? Consider the following warning signs.*

Signs of Complacency

Your group may be stalled by a cloud of complacency and apprehension. The negative attitudes can result in shutting down the flow of fresh ideas for preventing competitive attacks as well as for growing the business. Also, personnel may be overly preoccupied with defending an existing market position. Whereas, little time and effort are spent searching for new market and product opportunities.

Another obstacle is the negative workplace environment where other managers and staff are caught up by fear of what competitors might do. Such emotions discourage any effort to mount a vigorous response strategy in the event a competitor takes direct action that could threaten your strategies.

As an outcome of such an environment, you may find yourself exhibiting undue caution, which permeates the group and discourages them from reacting to threats, or from going after new market opportunities. The organization thereby loses momentum and is vulnerable to the internal and external conditions that preceded the original crisis.

Signs of Inflexibility During a Time of Disruptive Change

Inflexibility means rigidity and the inability to come up with back-up strategies to outmaneuver competitors and prevent a setback. Within that same framework, no plans exist to take advantage of timely opportunities to establish a foothold into an unserved or emerging market before competitors are likely to react.

---

* Remedies for the warning signs are detailed in the following chapters.

Signs of inflexibility also suggest that no dependable competitor intelligence is in place, whereby action plans and strategies can be rapidly implemented based on facts, not mere speculation. Allied with intelligence gathering is a benchmarking system that periodically evaluates strengths, weaknesses, or best practices that can be used to develop flexible strategies. Here's where a cross-functional team is highly useful to tap the diverse backgrounds of individuals and develop strategies to enter markets or defend against an aggressive competitor.

Signs of Lethargy

This sign is characterized by a general malaise that permeates the organization, which results in missed opportunities and excessive delays in reacting to time-sensitive market opportunities. All of which can result in losing market share, securing a favorable competitive position, and possibly upending your entire turnaround plan. In turn, lethargy parallels the conditions of inflexibility described above.

In such a state, personnel appear to lack initiative in implementing business plans with any urgency. They are slow in preventing a product from reaching a commodity status, a condition that often results in price wars.

Those symptoms can often be a sign of inadequately trained or inexperienced staff, which contribute to an absence of organizational vitality, thus hindering forward movement. In turn, that results in an inability to act boldly and rapidly. Another issue is excessive organizational layers that prolong deliberation and delay decisions, creating a trickle-down corporate culture of procrastination.

What follows is failure to secure a competitive lead due to sluggishness, such as failure to integrate technology, which can exacerbate the issue of bringing new products to the market fast enough to preempt competitors' moves. (Look, once again, at the Kodak case in Chapter 1.)

Signs of Unnecessary Dispersal of Resources

This sign is regarded as diffusing efforts by not focusing at points that have the greatest impact. It is a major failing in that it violates a primary principle of strategy—concentration—which permits a smaller organization to take on a larger rival with greater resources.

Dispersing assets without the guidance of the turnaround plan also means dissipating resources across too many segments thereby being strong nowhere and potentially vulnerable everywhere. (Concentration will be detailed in a later chapter.)

Signs of Inadequate Competitor Intelligence

This point is mentioned above, yet it is so vital to the development of strategies that further detail is required to emphasize its importance in developing a turnaround plan and securing its success.

The essential points are that without an early warning system, competitors' actions can frequently catch you by surprise and hamper your ability to respond with speed and effectiveness. This is especially the case where there is a lack of real-time information about market events, which can lead to indecisiveness.

In turn, that hesitancy can filter down and affect the attitudes and morale of your staff. Further, personnel, if left to form their own conclusions are inclined to misjudge, exaggerate, or underestimate the competitive situation without reliable backup intelligence.

In practice, a danger exists where there is a tendency to develop product launch plans without documenting market conditions, competitors' strengths and weaknesses, and buyers' specific needs. It is especially critical when accuracy is needed to locate an optimum market position and determine how to defend against a competitor's intrusion, and, thereby, avoid the possibility of another crisis.

Signs of an Anemic Corporate Culture

Corporate culture forms the backbone of your strategy. Culture shapes how your staff reacts to a competitive predicament and thus guides their reaction in the face of a crisis.

The telltale signs of a troublesome corporate culture include personnel operating within a closed-in and uninspiring work environment that prevents them from recognizing the hard-nosed inevitabilities of competitive threats. Such a reality exists where managers and key personnel fail to internalize the consequences of, for instance, falling behind in new technology. As worrisome, they don't recognize its potential for exploiting fresh opportunities and for sustaining a strong defense of its existing markets.

Thus, an anemic corporate culture is incompatible with the changing dynamics of the marketplace, which is revealed as personnel unresponsive to the subtle shifts in customers' buying behavior. It carries farther, where shaping competitive strategies that are aligned with a weak corporate culture jeopardizes the outcome of the overall turnaround plan.

Signs of Ineffectual Leadership

The primary attributes of competent leadership are insightfulness, straightforwardness, compassion, strictness, and boldness. Meshed with those qualities are decisions and actions consistent with the overall cultural values of the organization. Where they are lacking, the damaging effect can create dire circumstances for the plan, which can extend to the survival of the organization.

Some warning signs are when personnel detect a hint of mistrust in the leader's ability to assess market and competitive conditions, and in his or her ability to make timely and accurate strategy decisions. What could follow is that personnel begin showing signs of negative behavior, confusion, and unwillingness to take the initiative. This is especially apparent if they are rarely asked for input, and, if received, feedback is seldom offered.

Where that condition occurs, your organization or business unit is likely to be enveloped in a malaise that creates stress and anxiety, so that employees show fear of loss—loss of pride in the organization, loss of status, loss of the respect of their peers, or possibly loss of employment.

The physical signs that are missing within this environment are no formal channels for two-way communications, no interaction through cross-functional strategy teams, an ineffective system for recognition and rewards, and inadequate or irrelevant employee training.

Signs of Sagging Morale

Demoralized employee morale can be attributed to ineffectual leadership, as discussed above. Yet it forms such a powerful part in the drama of an organization's existence that it deserves its own special emphasis. Similar to some of the above signs, the damaging signs of a morale problem are listed below.

Employees visibly display a lack of respect toward leadership. They seem to feel insecure about their jobs and have doubts about management's concern for their well-being. It goes so far that some managers don't appear to be concerned about morale and their staff is seemingly left to fend for themselves.

Just as serious, there is no sign of a cohesive team spirit, which exhibits as a conspicuous lack of new ideas, innovations, or opportunities that reaches senior management. Then, there is the situation where employees exhibit erratic behavior; they seem discouraged, indifferent, fearful, and resist risk-taking.

This is apparent with the blatant absence of mutual respect, no workable two-way communications system, and lacking a formal procedure to encourage employees to submit innovative suggestions. Worse, yet, where new market initiatives and competitive encounters fail, management routinely lays blame primarily on poor employee performance, which further widens the gap in relationships.

Signs of Failure to Apply the Principles of Strategy

A pattern of thinking strategically, which encompasses time-tested strategy principles may be lacking among managers at all levels. Several strategy concepts have been accurately documented in the writings collected from antiquity through recent times.

Authors, academics, executives, historians, and strategists have identified sustainable theories, rules, and guidelines. Known as universals, the following list is divided into two categories: primary strategies and supporting strategies. They represent applicable concepts for a turnaround plan. Some were highlighted in the above listing of danger signs and will be discussed in greater detail in the following chapters.

Primary Strategies

Indirect strategies—Employing indirect strategies means that you deliberately avoid seeking a competitive confrontation. Instead, you recognize the ultimate aim in any campaign is possession of a territory or market, as measured by market share, dominant position, reputation, brand identification, or financial metrics.

To achieve that end, you attempt circuitous and surprise approaches, rather than those maneuvers that are direct, predictable, and readily anticipated by your rival. Even if a mere threat of confrontation is enough to cause the competitor to abandon his position, the indirect strategy has been achieved correctly.

Speed—The study of strategy will confirm that rarely is a case found of a clever operation that was prolonged. Procrastination is the enemy of success. So that speed is essential to the attacker, whereas delay is an advantage to the defender.

This is a core concept when launching a competitive campaign. In practice, the application of this strategy rule is to take advantage of the rival's unpreparedness. This is especially so where speed and surprise are the only means for gaining an advantage over a competitor.

Concentration at a decisive point—There is no higher and simpler rule of strategy than in keeping one's resources concentrated at a decisive point. It should represent the competitor's most vulnerable areas in which the consequences would have the greatest positive impact.

The aim of the strategy is to match your effort against the competitor's power of resistance, which is determined by the resources at his disposal and strength of his will. As for determining the competitor's assets, with proper competitor intelligence you can calculate the physical resources. However, measuring the strength of his will to resist your efforts is more difficult.

Strategic applications of resources—Two main ideas drive this concept. First, all available resources should be directed at the central objectives that make up the turnaround plan. Taking unnecessary detours from the plan's objectives wastes strength and resources, and is a violation of strategic thinking.

Second, as noted earlier, when dealing with a major crisis or any competitive campaign, it should not be viewed as an isolated incident. Therefore, all subsequent actions must be linked to the strategic plan, which should reflect a policy focused on the growth and development of a market, as well as on the strategic application of resources.

The offensive—Going on the offensive is a key strategy. Failing to do so, standing still, restrained by doubts and fears leads to acute problems. A central issue is a slide in employee morale and overall loss of momentum. So that, if confronted in a tough competitive campaign, rather than give in to indecision, it is far better to take some form of action.

Consequently, opportunities don't wait. A leader seizes openings without hesitation. Therefore, given the choice between the audacious and the most careful solution, the most decisive course of taking action is preferable.

Neutralize the competitor's capabilities—Neutralizing the rival's abilities must always be a dominant consideration. In practice, that means reducing or eliminating the competitor's advantages. That could be achieved by preempting the competitor's actions and matching product or service benefits, technology, pricing, and promotional claims.

It also could mean finding ways to make the confrontation more costly to him. In still another approach, neutralizing also can mean indirectly influencing the rival to exit the market completely by means of placing the competitor in a threatened or untenable position, which can result in a deep psychological impact.

The ending point of a campaign—The natural goal of all plans is the turning point at which the attack becomes defense. If one were to go beyond that point, it would not only be a useless effort, it could do nothing to add to success. In all probability, it would have a damaging effect. What matters, therefore, is to detect the culminating point with discriminative judgment.

Supporting Strategies

Circumvent strong points of resistance—Where confrontations are unavoidable, maneuver is used to circumvent concentrated areas of resistance. As described above, make use of indirect strategies to secure success by applying strength against a rival's weaknesses. Doing so permits you to take action on your own terms, rather than react to your competitor's actions.

Consider failure a temporary setback—This concept is more of a supporting attitude. More explicitly, it means not succumbing to the debilitating emotions of failure. To give life to a turnaround, any setback should be considered transitory. Doing so embraces a positive attitude that assumes alternative courses of action do exist for a turnaround. The new mindset stimulates the intuitive part of the mind to find innovative courses of action to think fresh options.

Use comparative analyses—Conducting comparative analyses (also known as net assessments) of internal and external factors before a competitor encounter is a fairly reliable predictor of the outcome. The analyses (aided with the availability of Big Data) have three aims: (1) find out the competitor's plans, (2) establish the decisive point, and (3) diffuse the rival's strategies. All told, it guides your plans, strategies, and creates numerous advantages toward a successful outcome.

Reduce friction—The turnaround plan on paper appears straightforward; the objectives are precise and the strategies are clearly stated. Yet, even the clearly defined plan is difficult to implement when pitted against the realities embedded in numerous internal and external events. The intruder is *friction*.

Friction appears in many forms, such as the leader attempting to motivate staff to action, yet members of the staff deliberately or by chance slow down the pace of the plan, or place obstacles in its path. In extreme cases, they boldly reject the plan through nonsupport.

Beyond the internal obstacles, friction exists at the market level consisting of numerous and unpredictable competitor, economic, and consumer conditions that can grind operations to a resource-draining halt.

Many of the above ideas appear in various guises, forms, and applications within diverse real-world competitive conditions. A case in point was the turbulent situation at IBM between 2012 and 2014. During that period, the company lost a bid for a major federal government project to build a cloud computing system for the CIA, valued at $600 million. It was a stunning loss to the technology giant that had served the U.S. government for decades.

An internal examination of IBM revealed that the company had been losing customers over an extended period of time to a prime competitor: Amazon.com. Amazon offered customers an alternative to Big Blue's outsourcing approach by offering a more simplified and economical rent-a-server-with-your-credit-card model. It's a system where organizations large and small can rent technology cheaply over the Internet instead of buying the types of costly fixed arrays recommended by IBM.

Externally, a further examination revealed it was a late entrant to the cloud computing market, which in part was due to the negative effect the cloud would have on its three core businesses: services, software, and hardware. As one business lost revenue, the other two were injured.

Yet, with those near-crisis conditions within a fast-moving technology world, IBM confidently viewed it as one episode in an otherwise long journey from the company's beginnings in 1901 as the Computing-Tabulating-Recording Company. The company evolved to the recognizable name of IBM and moved beyond selling accounting machines and punch card systems during the Great Depression.

That phase was followed by bold steps as it created and dominated mainframe computing in the 1960s. In 1981, it shaped an entire industry, which has had a monumental impact worldwide for decades with its introduction of the IBM PC.

All of these eventful phases jolted IBM by one crisis after another, which in each case was followed by a turnaround. Thus, the company benefited by an honored past, an illustrious history of accomplishments, and a set of organic values that combine into a wholesome corporate culture.

That culture reflects in excellent leadership that emerged during those tough times when Louis Gerstner, Jr. lifted the company out of near bankruptcy by conducting a brilliant turnaround. He was followed by Sam Palmisano who led his own timely turnaround by moving dramatically and successfully into software and services and away from personal computers.

Now, IBM is in a state of still another turnaround under the leadership of CEO Virginia Rometty who declared, "Our strategic imperative is absolutely that we will remake enterprise IT for the era of cloud. This is not the first time we've transformed. This will not be the last time." Embedded in those statements is her top strategy to reinvigorate the company's culture, especially during the current time of disruptive change.

Postcampaign strategies, then, represent the ongoing process that sustains the turnaround plan. Thus, you are able to more effectively focus on activating the strategies that secure the objectives of the plan.

As important, you can sort out the campaigns and more accurately identify the telltale signs that could trigger another setback. These could include events that created the original crisis and possibly still remain as overhanging clouds.

Consequently, you have to build defenses—both physical and psychological—to forestall further deterioration. Your essential task, then, is to take steps to lessen the risk of another crisis.

First, look for ways to apply your strengths against a competitor's weaknesses. The essence of the move is that you position your resources so that your rival cannot, will not, or simply lacks the capability and spirit to challenge your efforts.

Second, focus greater attention toward serving customers' needs and resolving their problems in a manner that visibly outperforms those of your competitors.

Third, search for a psychological advantage by creating an unbalancing effect in the rival manager's mind, whereby he or she hesitates in indecision. The aim is to disorient and unbalance the competing manager into wasting time and making irreversible mistakes.

Fourth, attempt to hold a long-term position in a target market, as gauged by attaining a market share objective, securing a position on the supply train, reaching a profitability goal, or similar metrics.

All of these steps require leadership techniques to activate change strategies, secure competitive advantage, and preserve success, which is the object of Section II that follows.

Section II

Activate Change Strategies

- Leadership Techniques to Activate a Turnaround
- The Competitive Campaign: Structure and Characteristics

4

Leadership Techniques to Activate a Turnaround

Chapter Objectives

Be able to

1. define how a leader's courage interfaces with the mind's mental power;
2. identify the primary traits of effective leadership;
3. describe how utilizing intuition can impact the decision-making process; and
4. adopt a mindset for oneself and the staff that failure is transient and only a temporary condition.

INTRODUCTION

A news story in a 2014 business publication revealed a case of an organization struggling with its U.S. sales. The company faced an increasingly competitive market that was changing rapidly. Quarterly earnings suffered its fifth straight decline in U. S. sales, which translated to a 5-percent drop in profits for the quarter. Also, according to the news report, competition from online retailers, such as Amazon.com was partly to blame for the slumping sales.

More distressing, however, was the accusation by a management consultant that the sales drop was a case of "self-sabotage with management's decision to cut worker hours and refusal to significantly raise workers' wages, which are largely the two big reasons why stores are increasingly sloppy and sparsely stocked." The organization reported on was Walmart.

A number of controllable and uncontrollable stumbling blocks also were hitting Walmart during that time period. Certainly changing buying patterns were sending shock waves through many retailers anchored primarily to the shopping malls, as buying patterns noticeably shifted to online purchases.

Then, there was the ongoing issue of a slowly recovering economy that was negatively affecting sales. Although economic issues were largely uncontrollable, organizations still had to develop approaches to operate under those adverse conditions.

Also during that period, Walmart's new CEO Doug McMillon entered and continued the ongoing turnaround efforts to reinvigorate the company's long-standing strategy of everyday low prices. He and President Bill Simon began making over its apparel departments to feature athletically inspired gear, accelerating the move to opening smaller stores like Walmart Express and Neighborhood Markets, and pushing rapidly into e-commerce.

But what can be said about the audacious claim of "sabotage?" That level of insinuation needs a response because any meaningful changes within Walmart require a cooperative and unified staff. The damaging claim focused directly on the issue of competent leadership and the ability of executives to tackle the problems of sagging morale, characterized by complacency, lethargy, and inflexibility. (Reports indicate the turnaround is well in motion.)

In this overall framework of activating a turnaround, what is the meaning of competent leadership? A continuing thread of leadership guidelines was woven into the topics covered on corporate culture, morale, and expectations from employees in Chapter 2.

Beyond these are deeper qualities that define effective leaders. First is the measure of intellect. Can leaders be labeled geniuses, icons, or legends, as portrayed by some enthusiastic journalists? If any of these terms describe leaders as successfully undertaking a complex set of activities and turning around a business with any degree of brilliance, then they can be legitimately referred to with these esteemed titles.

However, looking more closely into the realities of highly competitive marketplace conditions, similar to the type faced by Walmart, then an additional personal quality needs to be part of those terms: courage.

Leaders may be highly intelligent, but waver in doubt and indecisiveness. At times, they may be immobilized by insecure feelings, rather than activated by boldness, daring, and the determination to take action.

THE TRANSFORMING EFFECT OF COURAGE

Where implementing a successful turnaround as a measure of performance is at stake, and the characterization of a leader as a genius is in question, let's examine courage more closely. It is of two kinds.

First, there is a form of courage to withstand personal criticisms and attacks on his or her reputation and accomplishments. Second, there is a type of courage to willingly accept responsibility. That means an individual takes on responsibility fully aware of having to make challenging decisions that could affect scores, possibly thousands of employees, or having to choose strategies that could impact the future direction of an entire organization.

Courage may result from an individual's temperament or long-standing habit that can be considered a permanent trait. Alternatively, courage may result from such motives as ambition, reputation, and the quest for honors, which is usually attached to achieving a highly prized objective. This latter form of courage is usually associated with a feeling, an emotion, and not a perpetual state.

These two forms of courage act in different ways. The first is the more dependable and somewhat part of an individual's basic nature. It is more reliable and less likely to falter, and the tendency is for the mind to be calmer during hectic periods. The second form of courage tends to achieve more through spurts of boldness, which tend to stimulate others. The optimum kind of courage is a blend of both.

In no way does this emphasis on courage overlook the need for mental capacity. In all cases, the power of intellect is required to deal in the realm of uncertainty where sensitive and discriminating judgment is necessary to sort out the important from the trivial.

Notwithstanding the sophisticated intelligence programs available and the vast potential promised by utilizing Big Data, there still remains a chasm of information and assumptions that are open to doubt. With those uncertainties, chance is elevated to a high level, which entails substantial inputs of intellect, experience, and intuition. All of these are needed to develop and evaluate a new set of assumptions.

Therefore, when immersed in the everyday details of a competitive campaign, there may be no time to coolly review the situation, or to calmly think through all the possibilities. Usually new information and reevaluations are not enough to form a workable conclusion. If the leader is to emerge intact from the relentless pressure with the unforeseen, two qualities are indispensable:

1. An intellect that operates with some spark of an inner light that leads to correct decisions.
2. The courage to follow that faint light wherever it may lead.

The first quality is known as an inner sense; also known as intuition. To some people, intuition suggests an ethereal quality that can't be pinned down when it comes to developing actionable strategies and reducing the dire effects of unfavorable situations.

Yet, there is sufficient scientific evidence to confidently rely on this innate quality to take action and avoid being immobilized by doubt. This is especially so where rational thinking and sketchy market intelligence do not produce trustworthy solutions.

If you are able to experience intuitive assistance, then you can benefit from multiple impressions of instinct, insight, hunch, or gut feel. You also could receive sensations through such forms as vision, hearing, and perception. Intuition tends to be personal and takes on your inborn personality, while your mind goes to work on a problem.

For some strong-willed, in-charge managers, nothing can replace intuition. Even where managers lack originality, but display a determined personality, there are decisive moments when they must take counsel within themselves, make decisions, and move forward. Accordingly, trusting in intuition is a reliable leadership technique.

Seasoned managers consciously know the value of intuition in emergencies. However, they also are fully aware that intuition must be anchored to solid judgment and ongoing training, as well as to the concepts and principles suggested within these chapters.

Proof that your intuition is working appears when you reach a comfort level and where relatively sound decisions come almost automatically, so that you know intuitively, for instance, that one strategy is more likely to work, whereas another will fail.

To support decision making, savvy leaders also understand that many market events are more or less hidden in a fog of uncertainty. Such uncertainty is interplay of possibilities, probabilities, and good luck and bad that weaves its way throughout the process. Ambiguity is further magnified as they recognize that rival managers must rely on intuition as well. Moreover, those competing managers also are surrounded by dynamic physical and psychological forces that create damaging friction that cloud their judgment.

Therefore, your weightiest decisions are often made on inexact premises and there is always the element of chance hovering over all events. Yet, it would be totally false to assume that success is a matter of sheer luck. It is not luck in the ordinary sense that brings achievement. In the long run, so-called chance favors the courageous, intelligent, and intuitive manager.

Activating Intuition

Think of intuition as an inner voice that presents you with possible courses of action. It unlocks the mind and guides you by means of flashes of insights, ideas, images, or symbols. "The intuitive mind tells the rational thinking mind where to look next," declared the renowned Dr. Jonas Salk.

While intuition is often thought of as something elusive, spontaneous, and outside your control, nevertheless, it is possible to make intuition more accessible and more reliable. It begins by paying attention to your feelings. That means starting with a quiet mind and turning off the constant monologue that clutters the mind.

You then adopt a mindset where simply *knowing* transcends reason. Just perceiving possibilities is also an intuitive function. Of course, you can never be certain what the outcome of a decision is going to be. However, you can have a strong intuitive sense of the direction you want to pursue.

Still, such a leap of understanding isn't in opposition to or a substitute for reason, and it certainly is not in conflict with the input of reliable market intelligence and a system of knowledge management (Table 4.1).

Intuition is inner power that you use in addition to reason and factual information. In effect, there is an integration of intuition with the logical and linear thinking mind.

To make intuition work, you have to learn to hold attention on something for more than a few seconds, which can be a real challenge internally. For some, that also means emptying the mind of the emotional baggage

TABLE 4.1

The Power of Managing Knowledge

Explicit Knowledge	Tacit Knowledge
Exists in:	Exists in:
• Internal databases, documents, records • Websites • Analytics from call centers • Field services and reports • Marketing programs (data on consumer behavior and competitors' actions) • Back-office information	• Minds of individuals in the form of impressions, feelings, insights • Accumulated knowledge expressed through discovery and originated intuitively • Conversations, observations, and other types of informal interchanges over a gestating idea
Applications:	Applications:
• Determine internal strengths and weaknesses • Recognize changes in consumer behavior • Develop competitive strategies • Improve the accuracy of decisions • Justify expenditures for product development, implementing a fresh strategy, or probing an untapped market segment	• Transfers through interactions with others through a multidisciplinary framework for capturing, sharing, and disseminating information for developing new products and exploring new businesses • Prevents the loss of valuable and insightful information that may not be repeatable or retrievable • Assists in long-term strategic planning • Breaks down barriers among dissimilar business units

that upsets, angers, and creates a chaotic state of mind. It means letting go and being nonjudgmental.

You may have to personalize how you reach the receptive state where intuition can flourish. It may be through meditation, solitude, being out in nature alone, or using slow breathing techniques.

The object is to quiet the feelings, quiet the body, quiet the mind, and be left with a kind of inner knowing. In the silence, you learn the most about intuition. It is in this period of incubation that you let a problem just sit and you take time out.

Other approaches include using mental imagery, which is more associated with thinking and tends to be the type of intuition used by business executives, especially entrepreneurs, who tend to be highly intuitive.

In all methods, a common requirement is trusting yourself. Then you can trust your intuition. Keeping track of the accuracy of your intuitive decisions will give you an indication of whether your mind is prepped for intuition to thrive.

The benefits come when you can perceive possibilities in the future. In its most practical business application, intuition applies to developing a turnaround strategic plan where the first planning step is to develop a strategic direction for your company, business unit, or product line. Anything that is creative, breaks new ground, provides a future directed vision, or pushes you beyond the boundaries of what you already know is intuitive.

If any mystery exists about intuition, it is that people seem to get information, but they don't know how they got it. Mathematicians, for example, can arrive at theorems that have never been proved before, just through their intuition.

"No problem is solved by the same consciousness that created it," stated Albert Einstein, who was well-known for his use of intuition.

What is required is a shift in consciousness and, through this shift, you tap the subconscious mind to show the way to solve a problem, which is revealed as intuition. You then use logic, reason, and market research to follow up on the intuition for proof and validation. However, if time does not permit that luxury, then you must react and trust in your intuition.

The creative leap is always an intuitive one that enables you to see things that you haven't observed before. It is a new perception, as though intuitively you notice what you haven't noticed before, or you acknowledge what you already know but have forgotten. It is a sense of inner awareness rather than something that you need to learn about. Again, being intuitive is trusting yourself.

We are all equipped with intuition. It is built in. It is pragmatic. It solves problems. It identifies opportunities that may not be seen with conscious vision.

The Power of Determination

Add to courage, intellect, and intuition the power of determination. It surfaces when the mind informs the leader that boldness is required, and then provides firm direction to the individual's will. In short, determination originates from a type of mind that is strong rather than brilliant.

Often, determination shows most strongly among ambitious managers at the junior levels who strive to rise up the organizational ladder. However, once they arrive at the senior level, they frequently lose that special grit that propelled them upward in rank. At the senior levels, they tend to become increasingly conscious of the scope of their responsibilities.

They become more sensitive to the risks entailed by wrong decisions and the ramifications to the lives of personnel, the dangers of committing the organization's primary resources, and the possibilities of choosing a wrong long-term direction for a business unit or the entire organization. Where they once were used to fast-paced action, some leaders become more timid even as they consciously realize the dire effects of vacillating in indecision.

Notwithstanding such tendencies, determination is an essential quality of successful leaders, as expressed by the following individuals representing a cross section of personalities and disciplines:

> A dream doesn't become reality through magic; it takes sweat, determination, and hard work.
>
> **Colin Powell**

> I never could have done what I have done without the habits of punctuality, order, and diligence, without the determination to concentrate myself on one subject at a time.
>
> **Charles Dickens**

> Not only our future economic soundness, but the very soundness of our democratic institutions depends on the determination of our government to give employment to idle men.
>
> **Franklin D. Roosevelt**

> The truest wisdom is a resolute determination.
>
> **Napoleon Bonaparte**

> Let us not be content to wait and see what will happen, but give us the determination to make the right things happen.
>
> **Horace Mann**

> If your determination is fixed, I do not counsel you to despair. Few things are impossible to diligence and skill. Great works are performed not by strength, but perseverance.
>
> **Samuel Johnson**

Presence of Mind

Another leadership quality is the flexibility to deal with the unintended, unexpected, and the inevitable challenges related to a turnaround. Presence of mind calls for the ability to think quickly in the face of demanding problems that are bound to occur in any turnaround.

It infers speed and immediacy of action, which is usually an innate personality trait of the leader. However, it needs the hardening and molding that only experience and training can provide when steady nerves are required to respond vigorously and correctly to a crisis.

As long as employees move forward successfully the spirit and high morale, those qualities may not be put to their full test. However, what if the unintended challenges do occur and tough times prevail for some extended period during the turnaround? The leader needs tremendous willpower to overcome the negative effects on personnel.

One test of all these leadership attributes is the extent to which they impact the morale and physical strength of the staff. Ultimately, they are the ones who have to make a personal commitment of how much they are willing to give of themselves.

A second test is how effective the leader is in reviving his own sense of purpose and energizing his own morale before he can awaken the enthusiasm in others. Only to the degree that the leader can do this will he retain his hold on the staff and keep control. Once that hold is lost, once his own determination can no longer energize his employees, the effort will fail.

These are the burdens and obligations that the leader's collective qualities of presence of mind must overcome, if he/she hopes to achieve a successful turnaround, and the burdens increase along with the increases in responsibility.

Honor, Recognition, and Reputation

The qualities discussed above are not easily produced where there is no emotion or stimulus. Of all the passions that inspire individuals is the longing for honor, recognition, and reputation. There may be other compelling emotions, but rarely, if at all, are they a likely substitute for a desire for renown, which is driven by an individual's personal ambition.

As for honors, there is a seemingly insatiable demand for awards that is satisfied in the form of plaques, decorations, prizes, titles, and other tributes. There seems to be no limit to inventing new awards to recognize the manager of the month, of the year, by industry, by age, as in the outstanding leader under 40, and so on.

Psychologists have shown that people often value status above and beyond the monetary rewards. It reaches the point where such individuals are even willing to incur their own costs to "buy" high status. The object of these cravings is to satisfy the deep-seated human needs for social prominence, identity, recognition, self-esteem, reputation, and fame. The intent of these honors is to further inspire motivation, productivity, morale, and job satisfaction.

Psychologists further point out that individuals have an innate desire to distinguish themselves by harboring a strong urge to excel over others. Thus, the quest for social distinction is taken as a hardwired trait of human nature.

As for ambition, it is generally accepted that ambition is an essential quality of a leader. Some commentators make the argument that a successful leader is hard to find that doesn't possess a healthy level of ambition. However, for pragmatic application, ambition must be worthy of the organization's mission and not as a pathway to personal power.

From still another viewpoint, ambition is hard to separate from courage. In thinking about outstanding leaders, it is difficult to decide which of their actions in the face of severe problems bore the mark of boldness or that of ambition. Both are characteristics of the truly outstanding leader.

Constructive ambition, and the intense desire to excel, is the combination that stimulates ambition in others. Meaning that the magic of winning always arouses determination, which gives momentum to the organization. Thus, nurturing positive ambition in others, as well as in oneself, is an essential duty of the leader.

That said, the unwelcome reality also exists that unrestrained personal ambition does live on, with all its excesses and potentially harmful outcomes. It is uncontrolled raw ambition that can destroy employees' careers and the economic livelihood of the community in which the organization operates.

Such excesses and scandalous executive behavior within the financial community decimated even the loftiest organizations during the height of the 2008–2009 recession. Consequently, if you seek competence in leadership, understand how the power of ambition interfaces with the goals and culture of the organization.

Strength of Mind

This trait of leadership is shown as the ability to retain a sense of calmness and self-control during stressful periods. Strength of mind tempers the extremes of emotion when worry, anxiety, and fear are likely to emerge and cloud the reality of a situation, especially during the critical moments when rational thinking is needed. Ideally, it is rooted in an individual's temperament and will not be unbalanced by intrusive and negative emotions.

For instance, here is how individuals can differ in their emotional reactions:

1. Those individuals who exhibit a small capacity for being aroused. They are usually known for their calmness and lack of emotion.
2. Types of individuals who are extremely active, but whose feelings never rise above a certain level. They are known for their composure and sensitivity.
3. Individuals whose passions are easily inflamed. They are the ones who flare up suddenly, but soon burn out.
4. Those who do not react to minor matters and will be moved only gradually, not suddenly. When aroused, their emotional reactions will heighten in great strength over a longer period of time.

In the first group, unemotional individuals are hard to throw off balance. They are seldom strongly motivated, lack initiative, and consequently are not particularly active. On the positive side, however, they seldom make serious mistakes. Yet their lack of vigor cannot be totally viewed as being without strength of character. Within a narrow range, their self-control is useful during a turnaround campaign.

With the second group, some trivial matters can suddenly stir some individuals to act, whereas great issues are likely to overwhelm them. Within the group, one individual will likely help a fellow employee in trouble, but the problems of an entire group will only sadden him and will not stimulate the individual to do anything.

In the third group, there is no lack of energy or balance. However, those individuals are unlikely to achieve anything significant, unless they also possess a powerful intellect to provide the needed stimulus. However, it is rare to find this type of temperament combined with strong and independent minds. Their volatile emotions make it doubly hard for such individuals to preserve their balance. They often overreact, which is detrimental where stability is needed. That said, it is also untrue that highly excitable minds could never be strong and never keep their balance even under the greatest strain.

As for the fourth group, these individuals are difficult to move, yet they have strong feelings and are best to call on when there is a tough situation to overcome. Their emotions tend to change slowly. They are not overcome by their emotions, as is the previous group. At times, however, they too can lose their balance and be swept away by a highly emotional event. This can happen whenever they lack self-control.

The essential points to these characterizations are that leaders need the strength of mind to stick to their personal convictions, whether these are derived from their own opinions or someone else's, and whether they represent principles, attitudes, or come about from intuitive insights.

Such strength of mind won't show if leaders keep changing their minds, which is always possible through some external influences. In other instances, sudden changes may be the workings of their own intelligence, which also would suggest a curiously insecure mind.

In those cases where leaders' opinions are constantly changing, they are often viewed as indecisive, lacking vision, and hesitant to chase opportunities with vigor and conviction. These flashes of negative behavior are not easily hidden from members of the staff.

Employees can understand the need for flexibility where circumstances necessitate a change in direction, especially where the changes are communicated in a timely and logical manner. What they cannot understand is inconsistency and sudden erratic responses to situations that could be interpreted as strange and unstable behavior.

Strength of Character

Allied with strength of mind is strength of character. It, too, depends on leaders with balanced temperaments who can exhibit the emotional strengths and stability associated with a powerful character. However, strength of character at times can collapse into stubbornness. Although not the same, the line separating them is often hard to draw and should be judged on specific incidents.

Stubbornness, then, is not an intellectual defect. It comes about from unwillingness to admit that one is wrong. To attribute this to the mind would be an error. Rather, stubbornness is the fault of an individual's temperament.

Stubbornness may have its roots in a special kind of self-centeredness. In some ways, it can be likened to vanity or, in the extreme, a kind of obstinacy when suddenly and instinctively an individual rejects another's point of view, especially, without fair deliberation and reasonable thought.

In sum, strength of character is another way of saying: Remain calm and firm and avoid being easily unbalanced by negative events. This can be achieved by looking at the big picture and by focusing on the strategic direction of your turnaround business plan (see Chapter 3). In such a state, your mindset would tend to take on a more reasoned approach to any adverse internal situations or potential competitive and market challenges.

The whole object of strength of character is to sustain a sense of balance and avoid being humbled by the humiliating effects of anxiety, fear, and worry. If left unattended, those feelings can cause you to lose your grip on the situation and possibly end up losing the momentum of your turnaround.

How do these collective traits of leadership show up in the competitive world where all major factors are somewhat equal and leadership becomes the deciding factor in any campaign?

Consider the following examples from the 2014 Fortune 500 list:*

> **Jeff Bezos**, CEO of Amazon.com, continues to exert his visionary leadership with such bold plans as drone delivery of products to consumers and the Amazon Fire TV box to reach into the living rooms, which aims to actively take on Roku, Google Chromecast, and Apple TV.
>
> **Satya Nadella**, CEO of Microsoft, envisions turning around the company from its legacy platforms into a services business, which runs on data analytics, mobile devices, and cloud computing. His ambitious leadership approach is to ignite the drive for innovation among the workforce of 100,000 by recasting the mammoth organization as an underdog.
>
> **Mark Durcan**, CEO of Micron Technology, has outperformed about every other tech stock, rising by 212 percent in 2013 with its

---

* This author has no psychologically based insights or evidence about which of the leadership traits described in this chapter were dominant among the following individuals during the periods when they excelled. However, it would appear from their superior levels of accomplishment, several were strongly active.

computer memory chips, which consistently outpaced competitors. Strengthened by important acquisitions and aggressive strategies, Durcan has courageously positioned Micron to take on strong rival Samsung.

Larry Page, CEO of Google, takes his leadership thinking to reach far beyond its core business of search advertising into the world of robots, drones, self-driving cars, wearable computing devices, contact lenses that monitor a diabetic's glucose levels, and giant Internet service balloons.

Bob Iger, CEO of Disney, aims to infuse the organization with an entrepreneurial spirit by actively getting involved in the company's new startup accelerator. Beyond the continuing strategy of acquiring intellectual property, he is looking to the future by aggressively incorporating digital content through a multichannel network that runs a number of popular YouTube channels.

Mark Zuckerberg, CEO of Facebook, saw his company's stock climb sharply by 89 percent in 2013. His leadership is shaped by an excellent capability to think strategically, which reflects in a series of astute acquisitions, including Instagram, WhatsApp, and numerous others. Zuckerberg also pushes in-house innovation by unbundling Facebook's mobile app by launching Paper and Messenger.

LEADERSHIP APPLIED TO MARKET SELECTION

Up to this point, the attributes of a leader dealt with the qualities of mind and temperament working in balance. That concept now extends to the pragmatic issue of looking at the arena where the turnaround campaigns take shape: the marketplace. That is where market selection requires the full input of a leader's courage, determination, intellect, and the strengths of mind and character.

A useful way of categorizing markets is to divide them according to their essential features. You can select your strategies for entering, securing, and defending your position. These markets are identified as *natural, leading edge, key, linked, central, challenging, difficult,* and *encircled*.

Natural Markets

In this space, you are likely to operate in the familiar setting of ongoing established markets. The implication is that within such customary surroundings, personnel tend to be at ease and may not be motivated to venture out of their comfort zone.

Yet, to expand, you have to motivate them to move beyond the confines of existing markets before the mature, no-growth phase of the market life cycle sets in and you begin fighting a price war with competitors. That means you should get back to the issue of your organization's culture, as noted by the above individuals attempting to infuse an entrepreneurial spirit among their employees, and perhaps by acting as an "underdog." Your question then is: Does your organization's culture permit venturing out of familiar territory?

For the most part, in a natural market, you and your rivals have learned to adopt a live-and-let-live policy. That condition exits as long as each company sticks to its own dedicated segment. Generally, outright aggressive confrontations are seldom used.

The primary reason for this uncharacteristic display of togetherness in a highly competitive world is that you and your rivals share a common interest in furthering the long-term growth and prosperity of the market.

On the other hand, if any one competitor chooses to move forward and gain a meaningful benefit, likely strategies might include securing a more advantageous position on the supply chain by adjusting its position, or it can be done by adding or deleting a link in the distribution network, or it may choose to gain additional market share by recasting itself as a low-cost producer. Again, referring to the above examples, strategies could include acquisitions of companies, products, or technologies.

Leading Edge Markets

Leading edge means exploring markets by making minor penetrations into a competitor's territory. The intent is to investigate the possibility of opening additional revenue streams.

Such a movement out of a natural market requires that you obtain market intelligence to accurately determine the following:

- The feasibility for generating a revenue stream over the long term, and the possibility for expanding into additional niches.

- The investment needed to enter and gain a foothold in the market.
- A timeframe for payback and eventual profitability.
- An assessment of competitors: their market position, strengths/weaknesses, and nature of the opposition you will likely face.

A classic example of a leading edge market is the initial penetration by a few Japanese companies into North America with their small copier machines. Xerox, the market leader in those early years, concentrated its marketing efforts at large corporations with a line of large copiers.

Xerox managers initially avoided the small copier market. That oversight proved to be a critical error. Armed with a low-cost, no-frills desktop copier, enterprising Japanese copier makers moved in virtually unopposed and exploited a wide-open opportunity in the vast market of small- and midsize firms. Once established, they moved upscale in a segment-by-segment assault and took over a significant amount of Xerox's primary market share.

Key Markets

Key means that you and many of your competitors seem evenly matched within key market segments. The general behavior is that you would not openly create a conflict with an equally strong rival.

However, you may find that some competitor is attempting to dislodge you from a long-held position with the clear aim of taking away customers or disrupting your supply chain relationships. Then, you may be forced to launch a countereffort by concentrating as many resources as possible to blunt the effort. Such actions are appropriate, however, if they fit your overall strategic objectives.

Therefore, keep the big picture in mind. If you expend excessive resources in hawkish-style actions, such as price wars, then you may be left with a restricted budget to defend your overall market position. This condition is not what you want in a turnaround.

Linked Markets

In this category, you and your competitors are linked with easy access to markets. Your best strategy is to pay strict attention to constructing defensive barriers around those niches that you value most, and from which you can best defend your position.

Barriers include:

- Above-average quality
- Feature-loaded products
- First-class customer service
- Superior technical support
- Competitive pricing
- Generous warranties
- Patent protection of disruptive technologies

Not only do you build barriers against competitors' incursions, you benefit by strengthening customer relationships. In particular, customer loyalty gives you a long-lasting, profit-generating advantage that is difficult for a competitor to overcome. It is the one area that makes a meaningful addition to your growth. As one management analyst put it: "If you currently retain 70 percent of your customers and you start a program to improve that to 80 percent, you'll add an additional 10 percent to your growth rate."

Central Markets

Central relates to powerful forces that could threaten your market position. These forces are as diverse as watching competitors eagerly eating away at your position through aggressive pricing, or offering dazzling, feature-laden products, or introducing new applications through breakthrough technologies.

To counter such threats, look to various forms of joint ventures that can yield greater market advantages and offer more strategy options than you can achieve independently. For many companies, the merger and acquisition (M&A) route has proved the strategy of choice, as noted in the Micron, Disney, and Facebook examples.

Challenging Markets

In this category, if you enter a market dominated by a strong and aggressive competitor, be watchful. You could place your company at excessively high risk.

If, however, your long-term strategic objectives strongly support maintaining a presence in a challenging market, and, if the expenditures of financial, material, and human resources are consistent with your overall strategy, then find a secure position, for example, on the supply chain or in an emerging or poorly served niche.

Dell is a prime example of excellent supply chain management. In its original climb toward prominence, the company activated its manufacturing process and supply chain only when an order was received from a customer. That strategy eliminated the cost of storing excessive inventory. Dell benefited by shipping just the right amount of components to its factories, thereby avoiding investing in expensive warehousing.[*]

Difficult Markets

This type of market segment is characterized as one where progress is erratic and highly competitive. For instance, in attempting a meaningful market penetration and securing key accounts, or by maintaining reasonable levels of logistical support, you are likely to be blocked by asset-draining barriers.

If a competitor finds you off guard, and you subsequently lose your market position, it is difficult to return to your former position. In effect, you are entrapped in an untenable situation and your entire turnaround plan could be in jeopardy. Your best course of action is to go forward, only if the effort is consistent with your strategic direction and long-term objectives.

Encircled Markets

Encircled segments foretell a potentially risky situation. This market condition exists where you control limited resources and any aggressive action by a stronger, well-positioned competitor can force you to consider pulling out of a market.

In such a situation, it is in your best interest to maintain ongoing competitive intelligence, so that you can accurately assess the vulnerability of your position against that of your opponent.

Armed with the intelligence, you are able to develop a contingency plan that highlights your strengths and exposes your competitor's weaknesses. If, in your judgment, you still lack maneuverability and a capability to mount a meaningful competitive response, then exiting the market is prudent, as long as it minimizes disruption to your main line of business.

If, on the other hand, your competitor foresees an encircled position, it is wise to give the rival a way out of the market and not force him into a

---

[*] That strategy has worked remarkably well in past decades. However, as of 2013, with the decline in laptop computer sales, Dell changed direction and moved into corporate services and related products.

fight-to-the-end mindset. Your aim is to encourage him to take the more tempting approach and exit the market.

INTELLECTUAL STANDARDS AND PERFORMANCE

In summing up this chapter on leadership, it is useful to recognize that at various levels of the organization, and within each individual's area of responsibility, there are commonalities that include guidelines to a successful turnaround. These include:

1. A thorough grasp of the organization's culture
2. An awareness of the morale of those individuals involved in implementing the plan
3. Information about the availability of resources to achieve the goals
4. Updated intelligence about the market, including industry trends, customer behavior, and competitor activities

Then, there is a point where merging that information with the plan's strategic direction is used to shape objectives. From there, strategies are added to create action. Doing so gives the leader a greater awareness of the internal and external conditions. In turn, that process provides a realistic assessment of how much can be achieved with the means at his disposal.

How far into the organization can these turnaround guidelines spread? In its farthest reaches, that level of understanding would affect the salesperson within a specific territory. That individual would take on the strategic thinking and intellectual standards associated with a general manager and apply them to his or her sales area.*

---

* In this author's personal experience, several highly regarded global organizations provided training for its sales reps to assume the level of strategic thinking related to that of a general manager. In practice, it meant that each rep was required to prepare a strategic plan for his or her corner of the world.

Every plan included a strategic direction covering a five-year period (similar in format described in Chapter 3), long- and short-term objectives, strategies and tactics, and recommendations for new product and service development.

 Whereas the intellectual standards vary at each level of an organization, is it too great a leap to think that today's educated and intelligent workforce cannot think strategically, yet act tactically? This idea was also touched on in Chapter 1.

Finally, what sort of mind is most likely to display the qualities of excellent leadership? In the context of this chapter and with the central objective of this book focused on developing a turnaround plan, first, it is the inquiring, rather than the creative mind, the comprehensive rather than the specialized approach, the calm rather than the excitable leader, to whom the fate of employees and the future of the organization (or group) is entrusted.

Second, it is the leader who exhibits an advantageous balance of intellect and temperament. This includes the qualities of determination, firmness, strength of mind, and strength of character. Added to those are the individual's traits of ambition, courage, and tenacity.

Above all, it is the duty of the leader to train his/her own mind, and especially those he/she manages, to take on a mindset that failure is a transient incident; it is only a temporary condition. Armed with that attitude, the mind opens wide to release the creative powers to think of all the possibilities that could bring about a turnaround. In doing so, that mindset releases an inner knowing that the pathway leading to the objectives of the turnaround are altogether possible and achievable, so that the threats of failure, the roadblocks to progress, and the grinds associated with reversing a decline and forging ahead can be met and penetrated.

Of course, the realities also exist that some negative attitudes, within the leader and staff, are anchored to deep-rooted emotions. One workable remedy is to maintain the discipline by looking at the big picture, as defined by a well-developed strategic direction of the turnaround plan.

Thus, courage, determination, strength of mind, and strength of character are the essential underpinnings to maintain the mindset that keep you on course, bolster your staff during the discouraging moments, and prevent you from backtracking. Leadership as portrayed here is the starting point of all the events and the sustainer of all that follows. That step is always long from awareness to decision, from knowledge to ability, and from capability to action.

The following chapters deal with the essential components of campaigns and campaign strategy to create actions that lead to a turnaround.

5

The Competitive Campaign: Structure and Characteristics

Chapter Objectives

Be able to

1. recognize the causes of friction and how to reduce their damaging effects;
2. describe the essential components of a campaign;
3. distinguish between a defensive and offensive competitive campaign;
4. identify the termination point of a campaign; and
5. determine when to use reserves.

INTRODUCTION

Among the many factors that affect your ability to carry through with a turnaround is *friction*. This type of abrasive resistance takes many forms. A dominant one is the friction that prevails among staff members through their natural interactions and confrontations over status, preservation of power, or claims of authority.

These contact points include the adversarial relationships that often exist among diverse functions, such as marketing and finance, product design and manufacturing, sales and logistics, and the like. Then, there are the opposing personalities of individuals clashing from a variety of natural or contrived circumstances.

Another form of friction occurs in the competitive marketplace, as in the instance of Intuit, the financial software company that serves small businesses. With the average small business owner using as many as 18 apps to run his business every day, the data must flow seamlessly or, according to CEO Brad Smith, "it's going to become a point of friction."

To prevent this form of abrasive resistance taking hold and becoming a choke point, Smith opened Intuit's platform to allow any payment method to work with its products. Doing so permitted other developers to create products that would work on Intuit's platforms. With numerous competitors eyeing the market, such as possible entries from the likes of Google and Apple, Smith repeats with emphasis, "We can't be the point of friction in a small business's office."

One significant source of friction that should be anticipated comes about from expending excessive physical effort. Exhaustion is one of the characteristics that wears away at individuals' energies and, consequently, has an adverse effect on job performance, especially where extra levels of stamina and clear judgment are needed during the turnaround period.

Measuring the limits of this type of friction is difficult to gauge. Yet you can get a reliable indication through astute observation and a mind open to comments and gestures that reveal excessive levels of fatigue.

These signs are noticeable among those individuals who spend extensive periods of time in travel beyond their normal schedules. Or, it could be through excessive amounts of overtime and weekend work over long periods that show up as mental and physical wear and tear.

Thus, from a leader's point of view, expending physical energy for an extended period of time should be a major concern. What matters most is the effect of fatigue on staff morale, which could negatively affect the outcome of a campaign. It is the singular issue that makes a critical difference between success and failure in implementing the entire turnaround plan, especially where attitudes tend to be so fragile during the turnaround period and mood swings can deepen into depression.

Leaders respond in various ways to how their staffs react to the debilitating effects of using excessive amounts of physical energy, experiencing fatigue, and living with friction. Their overall approaches take into account such variables as attitudes, morale, and physical and mental well-being.

For instance, approaches include providing employees with multiple weeks or months of paid sabbaticals, rewards in the form of periodic profit-sharing bonuses, and professional assistance in planning personal and professional goals.

Then, there are other methods, such as helping individuals with personal finances, permitting participation on prestigious teams to develop new product ideas, allowing individuals greater freedom in making decisions that were previously reserved for more senior-level executives, or being selected to participate in special projects, such as improving workplace practices.

Market intelligence is still another source of friction. Specifically, that means acquiring tangible information about competitors, yet holding up on a decision when you know the intelligence is of questionable accuracy. The danger, then, is carelessly using the information when developing your own strategies.

The discomforting feeling is that intelligence is transient and can be as changeable as the minds of rival executives at any given moment. Thus, much competitor intelligence can be contradictory, even false, and certainly inexact.

Notwithstanding, even with the potential downside of faulty intelligence, there are substantial clues to help determine weaknesses and strengths of competitors and be able to pinpoint decisive points or segments on which to concentrate resources. Thus, to overcome this form of friction, level-headed judgment, intuition, experience, and a positive mindset are required when applying intelligence. The human factor still remains one of the all-important variables in decision making, especially where you have to deal with possibilities and probabilities more than with accurate facts.

What complicates matters further is the leader who is detached from the progress of the campaign and the goings-on in the marketplace. Or, where plans are developed by outside individuals who are even further removed from the action, as was brought out in the J.C. Penney case cited in Chapter 2.

Then, there is a matter of the very human trait that many individuals would rather believe bad news than good, and would tend to exaggerate unpleasant information. One remedy is for you to recognize this human foible and trust in your better judgment and experience, then convey hope rather than give credence to fears and doubts. Only then can you preserve a proper sense of balance that will keep you steady during the tough periods.

Balance is exactly what is needed when the real-time images and impressions of ongoing events appear entirely different when they actually occur. The senses tend to make a more vivid impression on the mind than systematic thought. This condition takes on more serious consequences when one executive replaces another, and passed-along impressions are communicated to influence the unaware successor. Thus, relying on your own senses and trusting on self-reliance is the best defense against false impressions.

In all, there is a great chasm between planning and implementing. On paper everything looks straightforward—the strategic options are prioritized, the product or service is ready to go, the tactical marketing campaign is ready to launch, and the logistics are in order.

Although seemingly straightforward in concept, the reality is that even the simplest things tend to become complicated. They accumulate and result in a kind of friction that is inconceivable unless you actually try to implement a plan. Then, the difficulties become clear.

They include countless minor incidents, the kinds one can never really foresee that combine to lower the general level of performance, so that one can easily fall behind reaching the intended objective. Of course, effective leadership, willpower, and determination can control various types of friction.

Friction, then, is the dominant force that more or less separates reality from words on paper. On the surface, the organization and everything related to it are basically clear-cut and, therefore, seem manageable. However, you need to bear in mind that none of the components consist of one piece. Each part is composed of individuals, every one of whom retains his or her potential for creating friction.

That is why discipline, effective communication, high morale, and adherence to a well-developed plan can contribute to welding the staff together for a unified effort. Consequently, the leader must be an individual of tested capabilities, so that the workings of the system move with a minimum of friction. The areas of friction mentioned above are only illustrations that could place strains on any turnaround and thereby make chance unavoidable and pervasive.

Contributing to the solution, in addition to forging discipline, are properly trained personnel skilled and flexible enough to eliminate some areas of friction. The unfailing lesson endures: Only the skilled, experienced, and trained will survive.

The following two organizations illustrate how these approaches are implemented.

Intuit, the company referred to earlier, creates 10 percent unstructured time for all of its 8,000 employees, so they can work on problems they see as causes of friction with customers. "We want to have an environment where the top talent can do the best work of their lives," declared CEO Brad Smith.

An organization that puts it all together is John Deere, producer of farm equipment. It tapped the experience of its workforce to implement the company's strategies. Teams of assembly line workers crisscrossed North America and talked to dealers and farmers about Deere equipment.

They traveled singly or in small groups and pitched their product stories to farmers at regional trade exhibits. Workers in various job functions routinely made unscheduled visits to local farmers to discuss their problems and needs.

In most places, the "new" reps were accepted as friendly, nonthreatening individuals who had no ulterior motives other than to present an honest, grassroots account of what goes into making a quality Deere product. At the time of initiating the strategy, enlisting the workforce for marketing-related duties was triggered by the weakening of demand for farm equipment during a recession, as well as by the aggressive actions of competition, in particular from Deere's chief rival, Caterpillar Inc.

Underlying the workforce strategy was Deere's view of customer loyalty: All employees are valuable resources to serve the needs of customers. Further, many of the workers supporting the effort had over 15 years of experience with the company.

They were trained in advanced manufacturing methods, total quality programs, and teamwork. According to Deere's management, harnessing that expertise demonstrated to customers that as makers of the products they were the best company spokespeople.

From Deere's viewpoint, a great deal was accomplished; customers' problems were identified early on, as well as likely threats from competitors. By working closely with customers, they uncovered potential new benefits that could be considered back at the home office.

Consequently, a major benefit was achieved by mobilizing the workforce to support its customer-loyalty efforts, which complemented Deere's core competencies, products, services, and cultural values. All of this information was internally communicated to deliver a powerful message of management–labor harmony, which kept in check possible areas of internal friction.

Altogether, the approach strengthened customer relationships by capitalizing on Deere's employees' experience, insight, and maturity. In turn, it resulted in a sharp increase in net income, along with sizable jumps in sales and market share over the following reporting period.

In sum, understanding the foundation concepts and underlying nuances associated with friction should be taken with the utmost seriousness. That means making every effort to internalize the full scope of its capacity for irreparably damaging your plan and then taking steps to reduce its impact.

THE ESSENTIAL COMPONENTS OF A CAMPAIGN

A turnaround most often consists of numerous individual campaigns, large and small, simultaneous or consecutive. They can be built around a sizeable market segment, which represents a major source of revenue, or it can be in a minor niche against a competitor, for instance, over sudden promotional discounts. Each campaign has its own purpose, which is subordinate to the main goal of neutralizing the rival's capabilities that prevent you from achieving your objectives.*

The main points are that the campaign is where the central action plays out as one organization contests the action of the other; that is, neutralizing the rival with the primary object of reducing the effectiveness of the other. All other organizational activities are purely in support of influencing the outcome of the turnaround.

Exactly what is meant by neutralizing the rival? It means taking actions that reduce the competitor's strength relative to one's own. These include setting in motion actions that cause the rival to spend excessive amounts of resources defending its market position, or it could be challenging the competitor's claims to product quality, features, or warranties. Your aim is to locate and exploit the competitor's weaknesses and vulnerabilities (see Chapter 1, Table 1.1).

Neutralizing also could take the form of carefully deploying salespeople in key territories, or by judiciously selecting intermediaries along the supply chain. If skillfully handled, the effects could unbalance and destabilize a rival. The result would be that a neutralized competitor would be at a competitive disadvantage and discontinue the campaign without incurring excessive losses, or, in the extreme, be forced to exit the market.

Of course, any of these actions should be undertaken without excessive expenditures by your own organization. If the rival exits the market or

---

* At this point, it would be useful for you to again review the numerous types of campaigns in Chapter 1.

retreats into a narrow segment of the market, at that point, the real profitable benefits accrue from increases in revenues and market share.

The above forms of neutralizing refer to the physical conditions where products, pricing, promotional, or supply chain strategies are the focal points of the campaigns. Then, there are actions that result in causing destabilizing effects on the morale of the opposing side, at the same time raising the morale on the winning end.

Consequently, in every campaign, look for signs of positive or negative changes in morale, courage, determination, confidence, and team cohesion. Any of those changes can contribute to the success or failure of your campaign.

Admittedly, such psychological and behavioral attitudes are difficult to interpret. Yet, it can be done by carefully watching how a competitor responds to your own actions. For instance, such telltale signs include noticing the speed of reaction, calculating the levels of commitment made in personnel and other resources, or observing if any of the aggressive actions even get a nodding response. These nuances of behavior provide clues to the level of morale about the competitor as well as your own.

An organization that continues to dazzle the marketplace with its spirited performance is Google. After going public in 2004, it swelled from 2,600 employees to 52,000 by 2014. Its quarterly revenue at the onset was barely $800 million compared with $16 billion for a recent period.

Today, the sprawling company continues to regenerate itself with innovative products and services as it explores new markets and futuristic types of businesses. CEO Larry Brin gives employees time off to search for the next thing that will push the curve of technology to new limits, from wind power to robots. Its current crop of competitors, such as Apple, Facebook, Amazon, and Microsoft have only to observe Google from any vantage point and get a clear picture of the company's culture, morale, and psychological makeup.

With the extensive references to morale here and in Chapter 2, there are specific ways you can create a morale advantage:

- Manage through availability and visibility. Show genuine interest by listening to employees' problems, complaints, and other issues.
- Manage with integrity and transparency. To the extent you can reveal sensitive information to personnel, explain management's future plans.

- Prioritize actions with input from others. Here is where the cross-functional team is useful to create a collaborative environment where individuals can share expertise in precise areas and where all can benefit from the diversity of opinions and backgrounds.
- Support a spirited, optimistic, and entrepreneurial work environment through ongoing training.
- Communicate often and openly, especially about market victories, both large and small.

In summing up the concept of neutralizing the competitor, there are three elements to consider: (1) the rival's loss of resources, (2) its loss of morale, and (3) the rival admitting to the losses and giving up its competitive intentions by exiting the campaign.

Duration of a Campaign

Because any competitive encounter means mutual loss through the expenditures of time, personnel, and physical and financial resources, much can be said about the importance of the size and duration of a campaign. The initial and obvious response is that the success of the campaign can't come soon enough for the winning side, or delayed long enough for the losing side, especially if there is a chance to buy time and regroup. (More on this topic in Chapter 9, The Ending Point.)

The reality is that rarely is a campaign decided in a single event, and, for most situations, the campaign shouldn't be given up, except where the objectives have changed. Or, there could be circumstances where the initial purpose of undertaking the campaign has shifted, morale has plummeted, or resources have run out and there is no sign of a new funding being approved.

It also is likely that leaders on both sides have already internalized the gravity of a serious situation well before it becomes apparent to others, and knows when to break off a campaign.

When is it the end? At least three factors should be taken under consideration before making a final decision to disengage from a campaign:

1. What effect will such a move have on the organization? Foremost is the ever-present consideration of employee morale. It is difficult enough to sustain it during so-called normal periods; to elevate it during a retreat strains even the strongest personalities. Then, there is the issue of available resources, and the personnel with the courage

and confidence to go back to senior management and make a convincing case for additional commitments.
2. What effects will there be on the rival organization? A downward slide in morale on one side is likely to cause uplifting feelings on the other, a sensation that only success can achieve. Where there is success, it is often followed with a building of momentum fortified with confidence that looks for further accomplishment. (See discussion on exploiting success, further on.)
3. What actual influence would disengagement from a campaign have on the future course of the turnaround plan? Here is where you take on a broader strategic view of a particular campaign and assess its contribution to a successful outcome. Is there an alternate time and place to face off with the competitor where the conditions are more in your favor?

If taking these factors into consideration and if convinced that continuing the effort is the correct decision, it may be necessary to revise portions of the plan and resubmit it to convince upper-level management of the strong possibilities of regrouping, recovering lost ground, and continuing with the odds that swinging to a success is a reality. Reaching such a critical juncture brings you back to the underpinnings of what makes effective leadership, discussed in Chapter 4: courage, determination, strength of mind, and strength of character.

Conducting the Campaign

No matter how a particular campaign is handled or what factors need special considerations, there are general accepted principles about campaigns that are worth noting:

1. Neutralizing the rival's capabilities is the overriding aim of a campaign. In turn, it dictates the specific objectives and types of strategies to use.
2. Neutralizing can usually be accomplished only by means of a campaign.
3. Only major campaigns involving all forces lead to major successes.[*]

---

[*] "All forces" means the traditional marketing mix: product, price, promotion, and supply train. That infers consolidating and utilizing all of these components into an optimum force.

4. It is necessary in a major campaign that the leader be directly involved, visible, and in control of the overall direction of the operation.

These statements lead to the following supporting concepts.

Neutralizing the competitor's capabilities is generally accomplished by means of significant campaigns where the results have meaningful effects on the total effort. However, this point doesn't negate the idea that a minor campaign to a greater or lesser extent can create a favorable outcome in that it could contribute to restricting a rival's capabilities.

That said, the major campaign is still to be regarded as the focal point of the plan's objectives and should not be avoided, as it is still the only real means to success. As a rule in a competitive environment, shrinking from a major decision by evading a campaign with a rival carries its own negative results; namely, it prevents coming to grips with necessary actions to complete the turnaround. If anything, avoiding it only prolongs the entire effort. Once again, morale is at stake and loss of momentum is a result.

If there is a delay, the usual reason is that the leader needs more time, which could work for or against achieving success. On one side, a postponement may have justifiable reasons, such as a product is not ready for delivery, logistics are not properly organized, and similar conditions. However, such a move should be carefully validated because the consequences could be serious.

If, however, the stall is due to a leader's hesitation triggered by doubt or fear, or where the winning spirit, courage, and determination that tip the scales are missing, more severe issues are at stake. And the following questions require answers:

Were there signs that the leader was unable to inspire members of the staff?
Can the same leader continue directing the campaign?
Was the negative behavior a temporary condition caused by some transient, and legitimate, incident?
Is the campaign recoverable and can that individual pick up the slack? What is the likelihood of such behavior repeating?

And additional questions need answers:

What was the actual purpose of the campaign in question?
How does it fit into the overall turnaround plan?

What are the conditions or signs that allow for a successful outcome and when is it likely to end?

There is a remaining consideration that is of utmost importance to a campaign: No campaign is complete without exploiting the initial success. As was pointed out earlier, this is where the real benefits of success are harvested. Therefore, where a competitor concedes, for instance, a market niche, that is also the point where consolidating the effort is called for with the aim of garnering more revenues, market share, or whatever metric of success is used.

What, then, is the problem that prevents such action? The leader's own energies and those of his staff may be sapped by mental and physical exertion. So, it can happen that for purely human reasons, less is achieved than is desirable or possible. Whatever follow-up does take place is purely due to the leader's personal ambition, energy, and ability to lead or drive his people forward.

Then, there is the consideration of how much time and resources will be expended in relation to how much will be gained. The importance of exploiting the first phase of the campaign is chiefly determined by the vigor with which the immediate pursuit is carried out. In other words, follow-up makes up the second phase of the success and, in many cases, is more important than the first.

Defense versus Offense

There is still another major concept that you need to consider about a campaign: The advantages and disadvantages of defending what already exists versus exploiting new markets.

The objects of defense are first to preserve one's market position against the inroads of competitors; second, maintain positive relationships with its customers; third, protect the organization's reputation, image, brand, as well as any other tangible and intangible objects or ideas that represent a sustainable and valuable resource for your organization over the long term. Should any of these areas reach the point of being in jeopardy, then the viability of the organization, group, product, or service could be in doubt, and so, too, the turnaround.

For instance, the J.C. Penney case discussed in Chapter 2 illustrates these objects. During its turnaround attempt, the CEO at that time failed to defend the core constituency that represented financial sustenance for Penney: its historic and loyal customer base.

What are the central lessons? It's axiomatic that retaining existing customers is far more profitable and easier than paying heavily to acquire new customers. Thus, setting aside enough resources to hold on to customers is a prudent strategy. It also is more advantageous to hold a well-entrenched position than to take a new one.

What follows, then, is that defense is advantageous to attack, assuming both sides have equal means. However, relying totally on defense would be completely contrary to the intent of a turnaround and growth over the long term. Thus, what is needed is an offensive component to every defense.

The central ideas, therefore, consist of two parts: (1) wait for the aggressive competitor to reveal its strategies, and (2) concentrate one's strength against the weakness of the rival, or on a segment that is worth targeting.

Waiting for the rival to show his hand has a subtle offensive aspect to it. While the notion is to observe and make every attempt to interrupt the competitor's moves, the intent is to cause the rival to exhaust itself through the expenditures of physical, financial, as well as any resources that would negatively affect the morale of its personnel.

That said, use defense as long as weakness compels. Then, abandon it as soon as there is a strong enough capability to actively pursue a positive objective.

When one has used defensive measures successfully, a more favorable balance of strength is usually created. Thus, the natural course in a turnaround is to begin defensively and end by taking the initiative. It would, therefore, contradict the very idea of a competitive confrontation to regard defense as its final purpose. The essential point is that one cannot attain any successful outcome from a campaign, nor achieve a turnaround, by staying totally on the defensive.

The classic case of Xerox, briefly mentioned in Chapter 4, illustrates this concept. Decades ago when Xerox created the market for xerography, the company initially focused on large companies with its large copiers. Consequently, its management focused on defending its presence in that market, but did so through passive resistance. That is, it left exposed a vast market of small- and midsize companies for small, tabletop copiers.

Astute Japanese makers of copiers, such as Canon, Sharp, and Ricoh saw the opening and attacked that vacant market without opposition. Once secured with a solid foothold in North America, they made the next expansive move of going upscale where they confronted Xerox head-on in its big copier stronghold.

That scene occurred in the 1970s at the introductory stages of the industry and product life cycles. It is conceivable that Xerox could be excused because various user applications were not fully explored at the time, nor had all user segments surfaced.

Certainly, to the company's credit, Xerox defended its position over the following decades and successfully counterattacked to recover a good deal of its market share. Even today, "make a Xerox copy" remains a generally accepted term for product identification, regardless of the brand of copier.

Thus, defending one's territory is certainly a prudent strategy. However, it can't be an inactive one, it must be actively defended, which means activating market intelligence to reveal indirect pathways for rivals to enter and threaten one's core market.

The Characteristics of Offense

Offensive action is generally complete in itself. That is, movement is always present; it doesn't have to be complemented by defense. However, practical realities intervene at various times, e.g., budgets dry up, management shifts attention to exploit new opportunities, competitive conditions force the decision to defend losses elsewhere, or personnel are transferred to new areas and no replacements are forthcoming.

As indicated earlier, there is the very real issue of employee fatigue, especially when an offensive drags out over an extended period of time, which is followed by declining morale and inertia taking over. At that point, the entire campaign is at risk, and perhaps the entire turnaround as well.

Consequently, the prudent approach is to defend what has been achieved. In practice, the offense becomes a constant interchange and combination of offense and defense. To take the concept one step farther, every offensive action will likely end in a defense whose nature will be decided by any or all of the circumstances described above. It follows, then, that every offensive move has to take into account shifting to the defense, followed again by the offensive.

For instance, a news report (at the time of this writing) indicated that Amazon was readying itself to attack Google's online ad business and take a piece of that multibillion dollar market. The online retailer reportedly was building a new ad platform similar to Google's. Known as Amazon Sponsored Links, it would be similar to Google's AdWords product.

Amazon's attack could be a significant blow to Google's bottom line, as Amazon is one of Google's largest ad buyers. With online ads accounting for the major share of Google's revenues, Google's offensive would focus more strongly on the defense.

A number of salient points can be summed up about the offensive. An organization on the offensive should have superior strength to make up for all the initial advantages that naturally accrue to the defender. That is why concentrating resources at a defender's decisive point is the preferred strategy for the offensive.

Also, offensive moves can use surprise to unbalance the defender. This is a potentially potent force with excellent possibilities where the defender has a cumbersome organization with layers of management that prevents quick communications from the field to decision-making executives.

Surprise takes on a more meaningful quality when considering how the defender's morale would be affected. Thus, the lower the defender's morale, the more daring the attacker could be.

Maneuver and diversion are still other characteristics that are most often associated with the offensive. Both aim to bring about favorable conditions for success. Such movements intend, first, to force the competitor to scatter its resources, although they have little effect on altering the circumstances of the campaign; second, to identify where to concentrate next efforts, while preserving one's own resources.

Campaign Follow-Up

For total success, an offensive campaign depends on the availability of resources. The problem, as indicated earlier, is that over time resources are consumed. Then, at one point, every advance reverts into a defense, as portions of needed resources are redirected to defend what was previously secured.

Whereas the natural tendency is to maintain the momentum and continue the campaign, it is at that juncture competitive conditions could decidedly change and a recharged rival cancels out your hard-won gains. The reason is that, as the physical and psychological strength of one side weakens, it strengthens the other side. Therefore, determining when to terminate a campaign takes discriminating judgment.

Even where one side is motivated by success, continuing an offensive campaign often risks the primary reason why the campaign began in the first place. Attention, then, should be given to the viewpoint that superior strength is not the end but only the means. The end is to neutralize the competitor, increase a position in a market segment, or achieve any other predetermined objective of the campaign. The object is not to fight for fighting sake and exhaust the company.

The key point? It would be useless to overshoot your target. You risk jeopardizing your gains and possibly forfeiting what was gained. It also could affect plans for shaping strategies and committing your organization's resources in campaigns that involve expanding into new markets against entrenched competitors.

Therefore, your task is to determine what conditions would optimally end a campaign. That is, would continuing the offensive and expending more resources place the entire turnaround in doubt and result in exhausting financial, material, and personnel resources?

The end point can be determined by quantitative calculations, compiling a list of nonquantitative criteria, or assessing various opportunities highlighted in your turnaround plan. An answer may come to you intuitively and turn out to be the correct one.

While the quantitative calculations can be handled with a good measure of accuracy, the nonquantitative ones are more judgmental and represent an important part of the decision-making process. Along with observing the attitudes of the staff and monitoring their performance, use the following guidelines:

1. Does the staff still display a proactive mindset when faced with a continuing and aggressive follow-up campaign?
2. How is the level of morale at the beginning of the campaign compared to the present point?
3. How does the staff react to risk-taking?
4. Does the staff require further training? In what areas and with what skills?
5. In there any sign of complacency and apprehension at various organizational levels?
6. To what extent is there active participation in the form of meaningful ideas and suggestions?
7. Are many individuals preoccupied with defending existing market positions, while giving negligible amounts of time and effort to thinking of advancing?
8. Are some managers and staff members caught up by fear of what competitors might do?

9. Is there an excessive display of negative behavior among key members of the staff toward mounting a vigorous pursuit in the event a competitor chooses to respond aggressively?
10. Are there signs of upper-level managers exhibiting undue caution, which may permeate the group and prevent pushing forward?
11. Is there any sign the organization or group is losing momentum?
12. Are there signs of hasty and erratic behavior, such as inclinations to misjudge, exaggerate, or underestimate a competitor's situation, as well as one's own condition?
13. Is the company's culture compatible with the changing dynamics of the campaigns up to this point?
14. What are the staff's attitudes toward the leaders and their ability to accomplish the turnaround?

Use of Reserves

The final characteristic of a campaign deals with the use of reserves, which is an extension of campaign follow-up. A reserve has three distinct purposes:

1. To prolong an active campaign against a competitor where market intelligence reveals that the rival doesn't have the wherewithal to hold out against a lengthy and sustained effort.
2. To be prepared to rapidly counter any unforeseen threats resulting from a surprise maneuver by a competitor, such as with a sudden promotional burst, price attack, introduction of a new after-sales service, and the like.
3. To exploit any breakthrough opportunity that appears in the form of emerging, overlooked, or poorly served market niches. Breakthrough also applies to a weakness detected in the rival's defenses, which can be exploited.

Within that framework, portions of an operating budget are held in reserve to accelerate the rapid adoption of a product with additional targeted promotions or by offering timely incentives within the supply chain. It also can be used to launch a new product feature at an opportune time.

In addition to the above purposes, a singular one stands out as a primary application of a reserve. It is for a time and place that represents a decisive stage in a campaign where a full concentration of available resources would assure a successful outcome in a campaign.

You can look at the purpose of reserves from still another vantage point. Maintaining a reserve should not be viewed as speculative with the idea that it may not be utilized. Instead, a reserve must be considered a strategic component of a campaign that would be used to decisively contribute to the overall success of the campaign. Consequently, the use of a reserve requires a careful assessment of where strategically and tactically it would result in the campaign's success. Otherwise, it becomes a meaningless resource.

As a rule, then, the final outcome of a campaign can turn decidedly in your favor based on the amount and availability of reserves. As long as you maintain more reserves than your competitor, the advantage is physically and psychologically on your side.

Should, however, your reserves start to dwindle compared to those of your rival, then failure becomes a strong possibility and would necessitate reexamining your entire turnaround strategy. In that situation, your follow-on moves would depend on what is going on in the marketplace, such as described above in the three purposes of reserves.

How you arrive at a correct comparative estimate of reserves with a competitor is a result of competitor intelligence and accurate internal assessments of your staff's morale and energy. To a great extent, it also would depend on your personal attitude, courage, and level of confidence, as well as the depth of your own skill and experience that you bring to the decision. Even, then, you can do no more than broadly influence the ultimate decision, which will be triggered by immediate marketplace and competitive considerations.

Finally, no matter how small the chance of success may be, if you hesitate to commit reserves, you will face the unfortunate point beyond where even determination can turn into faulty irrational moves.

Having discussed the root causes that trigger a turnaround, how to prepare the organization for a turnaround, the design of a turnaround plan, leadership techniques to activate a turnaround, and the structure and characteristics of a competitive campaign, the chapters ahead deal with the various components of strategy to activate the turnaround.

As the late management scholar Peter Drucker declared, "All plans must deteriorate into action." Action in the context of a turnaround is succinctly defined here as strategies to achieve the objectives of the plan.

Section III

The Essential Elements of Turnaround Strategies

- Bold Action versus Cautious Restraint
- Concentration versus Dispersal Strategy
- Indirect versus Direct Strategy
- Valuing Surprise and Speed

6

Bold Action versus Cautious Restraint

Chapter Objectives

Be able to

1. indicate the pros and cons of bold action versus cautious restraint;
2. cite examples of applying boldness to a competitive campaign;
3. identify conditions whereby caution would be justified;
4. list the types of competitive intelligence that would be needed when planning bold moves against rivals; and
5. describe the major features and benefits of each of the classic management tools.

INTRODUCTION

Courage, as an essential attribute of leadership, was emphasized in Chapter 4. That characteristic also applies to boldness, especially when it comes to neutralizing a competitor's capabilities. It incorporates the supporting traits of confidence, self-assuredness, and enterprise. So that, wherever boldness is exhibited by a leader, he or she is in a superior position to look forward, overcome fear, and expand positive expectations.

Boldness is required under the tough conditions of halting declining revenues and climbing back up to successful performance. With those grandiose outlooks, it is entirely reasonable to think of boldness as a creative force. So that, whenever boldness confronts excessive caution, boldness is the likely winner. Faintheartedness in a leader implies a loss of nerve and equilibrium, whereas the mindset should be directed to react with the sprinting thrust of a runner at the optimum moment.

Yet, reality does surface. As managers climb higher in the corporate chain where strategic decisions are made, an added dose of boldness is required. Instead, many who reach the C-suite tend to rely to an increasing degree on the conservative leanings of the mind.

It is there that they feel the full weight of responsibilities taking hold, and gnawing questions begin to form. Is the company's long-term direction correct or should a change be made? What about the lives of employees if plans fall short? Are there sufficient resources available to move forward? Is the organization sufficiently responsive to customers' needs and how can a change in mindset take place? Can personnel deal effectively with the threats from aggressive competitors? Are we better off just protecting the existing business and avoiding any notion of risking an expansion into new areas?

Under that mountain of inhibiting questions, boldness, which is also a quality of temperament, will tend to be consciously restrained. Where there are exceptions, the few outstanding leaders appear, such as the likes of Bill Gates, Jeffrey Bezos, Stephen Jobs, Larry Page and Sergey Brin, Mark Spielberg, and Jack Welch. They are the ones who break through the heavy barriers that envelop mediocrity.

What sets boldness apart among that breed of leaders is that they are imbued with superior intellect, insight, and the compelling need to make things happen. The stronger the combination, the greater is their reach for changing possibilities into opportunities and then into action. However, they realize that there is also a greater level of risk.

Whereas the average managers may arrive at correct courses of action, they are intimidated by the potential dangers of losing nerve, the willingness to take on risk, and the lack of a long-term perspective. Even if sound advice is provided by others, such individuals can lose their powers of decision and suffer the consequences of wavering in indecisiveness. Thus, in every practical sense, an outstanding leader without boldness is problematic.

Boldness, then, is a force bound by a leader's self-control. If disciplined, the leader can make decisions to commit resources based on the objectives of the turnaround plan, and not dissipate them into inconclusive activities, such as running after some unsubstantiated lure of a Monday-morning headline that in itself is short-lived.

There are still other dimensions that relate to boldness. An organization may be infused with boldness for three reasons.

First, it may be staffed by particular types of aggressive individuals who are deliberately recruited with that quality. One organization in the electronics field sought out individuals as marketing managers, product managers, and sales positions who had active military service and graduated from one of the military academies, such as West Point, the Naval Academy, or Air Force Academy. Beyond those initial credentials, only those who served as air force and navy fighter pilots, combat marines, and submariners were hired.

Second, it may be the result of seeking bold leaders within its ranks. They are the ones who take on the responsibility to educate their personnel in the spirit of boldness to counteract the insipid tendencies of complacency and lethargy, which taint people in times when energy and determination are needed. As an example, Google executives build teams known as "smart creatives" who are characterized as impatient, outspoken risk-takers.

Third, senior-level management makes certain there is an appropriate alignment of its corporate culture to foster boldness. Such a corporate culture subsequently drives bold decisions and inspires individuals to act with a corresponding mentality. That type of culture is again illustrated by Google where fast decision making and flat organizational models are a corporate way of life.

What does boldness look like in an organization going through a turnaround?

Consider Winnebago Industries, the maker of recreational vehicles (RVs). In 2009, the company looked as if it was going to close its doors. Sales had been declining for several years and fell into a sharp drop after the global recession of that year.

By 2014, however, it pulled off a bold turnaround, successfully outperformed most of the industry's surviving competitors, and saw its sales come barreling back. The big win was to see Winnebago on *Fortune's* list of Fastest-Growing Companies for a second year in a row.

Behind the turnaround was CEO Randy Potts. His confident moves were demonstrated by positioning Winnebago to survive the powerful economic forces that created havoc among his primary segments of retired consumers. Throughout his struggle, he grasped the opportunity to reach new levels of growth by firmly latching on to the demographic phenomenon that America's 76 million baby boomers were beginning to retire.

This vast group represented the primary audience for its RVs. Potts readied Winnebago with a new line of motor homes that had all the luxury features suited to a new generation of retirees: flat-screen TVs, a bunk with power lift, a Murphy-style sofa bed that converts to a dining table, tastefully colored cabinetry, and the latest technologies.

As the direction of the turnaround firmly solidified into a huge opportunity, the Iowa-based company forged ahead with bold promotions, from a dazzling Las Vegas product launch directed at dealers to lively big rallies across the country to engage its customer base to a variety of local events to involve still other emerging customer segments.

Within the organization, Potts also made other forward-looking moves that challenged and, subsequently, ignited the more laid-back corporate culture that had driven the company in past decades. He placed product managers rather than engineers in charge of making key product decisions for new vehicle designs. The result was a series of new vehicle introductions, including new models at lower price points that better suited the postrecession marketplace, such as its new generation camper that targeted younger customers.

In sharp contrast with Potts' bold performance, other executives, for a variety of psychological reasons, purposely avoided those types of moves at the time of decision making. Some are due to endemic personality concerns, such as periods of low self-esteem, which results in indecisiveness during tense phases of a campaign.

Notwithstanding that such individuals have somehow risen through the ranks, it's not uncommon for negative feelings to persist, even though they have learned to adapt successfully to the pressures of the moment and are able to act assertively when pushed for a solution.

There is still another issue related to boldness that is triggered by the innate fear of failure. An audacious move is an act of determination in a specific situation. It becomes a character trait only if it becomes a mental habit. There are ample numbers of brilliant executives who simply don't have what it takes, and they don't fully recognize that successful campaigns are waged under bold leadership.

There are some remedies to overcome these negative effects: First, the mere act of taking some action will often arouse the inner feelings of boldness. That is, it will help to push aside the critical moments and awful emotions that can creep into the mind and take control of natural movements.

A second remedy is to back up decisions with as much credible market intelligence as possible. Doing so, results in thinking more confidently about taking alternative courses of action. Such an approach helps support decisions, counters the feelings of indecisiveness, and results in clear-thinking actions at any given moment.

A third approach is to gain strength from the secure underpinnings of training, internal discipline, and the years of valuable experience to help overcome negativity.

These remedies, plus your natural intuition, would kick in to overcome any pessimistic emotional state. Keep in mind, too, that a leader is in a continuing contest of mind against mind—his mind pitted against the mind of a competing manager who may be challenged by similar emotions. It is often the self-assured one who prevails and moves forward.

Once again, consider the extraordinary success of Google Inc. Much has been written about the brash approaches that founders Larry Page and Sergey Brin exhibited when they started operations in 2000. Even then, they showed their boldness by going against the sage advice of seasoned consultants and analysts who advised selling out for a mere pittance during those early days.

Instead, industry pundits watched their daring surface again as the search giant quietly acquired a mobile phone software company and began making forays into the instant messaging, Wi-Fi Internet, as the company established its Android software for phones and tablets. And beyond, Google is pushing forward with its self-driving cars and other breakthroughs.

APPLYING BOLDNESS

Turning again to Winnebago, what strategic and tactical objectives would CEO Potts and his managers have likely considered for sustaining the company's momentum and countering any aggressive competitive actions?

First, he would certainly have listed specific objectives related to his core customer group of retirees. Then, he would undoubtedly have specified additional ones to establish a brand preference with the baby boomer segment.

Second, to assure that his plan is as secure as possible against sudden competitive attacks that would interfere with his plans, he would aim to neutralize his rival's capabilities.*

---

* See Chapter 2 for a list of various types of competitive campaigns.

Neutralizing takes many forms, such as misleading the competitor about the campaign's real intentions. For instance, it could include broadcasting false signals, within legal and ethical bounds, that would result in inducing the opposing manager to disperse his efforts in anticipation of where the next move would occur.

In effect, it would place the opposing manager on "the horns of a dilemma." The consequences of fear, apprehension, or confusion can cause any rival manager to reduce his presence through attrition or by exiting the market entirely.

There also are neutralizing campaigns designed to make staying in the market more costly for the competitor. This approach is possible if intelligence indicates that the rival cannot sustain lengthy campaigns, especially where heavy investments are required just to remain competitive and its financial condition is fragile. That type of campaign also includes being able to withstand lengthy price battles that would further jeopardize its finances and increase losses.

In some cases, depending on resources and perceived strength, just the mere presence of appearing formidable can be intimidating enough to frighten off one side or the other. It can exist in a relatively small market niche, where even a numerically small effort can result in a successful outcome, and it is possible in a larger market segment where greater resources can apply.

The essential point is that useable competitor intelligence is needed to make the appropriate calculations and estimates for follow-on decisions. At times, however, some managers tend to neglect this area. They may be constrained by time, or it could be the nature of their personalities to casually accept meager or inconclusive intelligence, because of their anxiousness to take immediate action.

Yet, making proper estimates becomes the basis for effective decision making, especially where it comes to planning how and when to conduct a competitive campaign, regardless of whether it is defensive or offensive. It factors in when calculating how much resources should be committed and in what form.*

What, then, is the kind of intelligence Potts should be looking for when making calculations and estimates?

---

* Chapter 5, The Competitive Campaign: Structure and Characteristics, reviews in detail defensive versus offensive campaigns and the use of reserves.

TABLE 6.1

Assessing Competitive Operating Patterns

1. Competitor's size	Classify by market share, growth rate, profitability, as well as any other metrics meaningful to your company and industry
2. Competitor's objectives	Uncover competitor's intentions related to product innovation, market leadership, global reach, regional distribution, and similar areas that would indicate a strategic direction
3. Competitor's strategies	Analyze rival's internal strategies (speed of product development, manufacturing capabilities, delivery, and marketing expertise), and form an opinion about its external strategies (supply chain network, field support, market coverage, and aggressiveness in defending or building market share)
4. Competitor's organization	Report on its organizational design, culture, operating systems, internal communications, leadership, and overall caliber of personnel
5. Competitor's cost structure	Check up on how efficiently it can compete, how long it can sustain pricing competition, the cost or difficulty of exiting a market, and its views about short-term versus long-term profitability
6. Competitor's strengths and weaknesses	Identify decisive areas vulnerable to a concentrated effort as well as those strong areas that should be avoided

First and foremost, he needs a clear outline of the competitor's operating patterns with as much accuracy as possible. That information would be essential if Potts is to uncover the strengths and weaknesses of his competitor's position. In turn, such information would prove valuable for determining the types of strategies to handle an offensive or defensive campaign.

Also, with some measure of reliability, he could then set reliable budgets and determine the form in which resources would be allocated. That would include marketing allocations directed for a push strategy through RV dealers or a pull strategy direct to end users.

Also, armed with such knowledge, Potts would be in a better position to motivate his staff to handle potential threats and opportunities. Table 6.1* provides guidelines for determining types of information useful for making estimates.

---

* Some of the intelligence in the six sections is more qualitative than quantitative and thereby subject to some speculation, rather than factual information. Nonetheless, the mere process of piecing together scattered pieces of knowledge can result in useable information and form an overall picture of the competitor's operating patterns.

Competitor intelligence, as an information-gathering and decision-making tool, affects all operating parts of a business either directly or indirectly. Obviously, for Potts to obtain useable intelligence should be a rather familiar process. The primary reasons for mentioning it here is only to highlight its two-sided effect when developing a bold competitive move.

First, it is not foolproof. Not only is the information in a continuing state of change, but once received by a manager the interpretation may be flawed depending on the personality, experience, and skill of the individual making the assessments.

Second, even with all the potentials for errors, competitor intelligence is an indispensable and intrinsic part of every offensive and defensive effort where developing strategy is at stake. It also has been an ongoing theme throughout the previous chapters and will be selectively mentioned in those chapters that follow.

The primary point is that Potts should acquire the best intelligence available. Then, it should be validated for the reliability of each source. Reports can then be confirmed or disproved with some reliability. Potts and his managers, thus, would gain confidence and become more adept at making accurate decisions. Intelligence, then, is the centerpiece of all offensive and defensive actions by exposing opportunities and threats for Winnebago.

One more piece should be added to intelligence gathering that requires Potts' attention: a reliable system for observing, listening, and then interweaving the numerous market signals that reveal substantive information about his competition.

The competitive marketplace pulsates with competitive actions with which he and other Winnebago managers can tune in to gain invaluable insights to enhance the accuracy of their decision making.

For instance, there may be strong signals about competitors opening or closing regional offices or plants that would directly impact Potts' operation, or there could be more subtle ones about sudden management changes. Other signals may reveal impending layoffs or rumors about new competitive alliances.

Not to be overlooked are the telltale signs of internal disorder and signals of inept leadership. These show as competing managers visibly appear discouraged, display low morale, or exhibit short tempers. Sales reps overtly look for other jobs, criticize working conditions, or complain about shortages of sales aids and supplies to anyone who listens.

They whisper about ineffectual leadership and cuts in wages. They whine about stringent rules restricting travel-related expenses, or they object to executives excusing (or overlooking) abuses in corporate procedures.

At the same time, there are understated signs of general disorder, sloppiness, or indications of internal anxiety. For instance, there may be signals that represent changes in a competitor's traditional operating style, or in patterns of handling customer problems, or in the general demeanor of how executives interact with their personnel and its subsequent effect on morale.

Other signals, from customers, might openly complain about the competitor's policies, rules, and procedures, and, if pumped for detailed information, they often surge forward with a flood of grievances.

All of these words and actions could indicate fear, uncertainty, insecurity, and a variety of deep-seated internal problems. It is in such a state of unbalance and discontinuity that Potts would spot opportunities.

Additional market signals could have implications for Winnebago, if one or more of the following occurred:

- A competitor abruptly announces a new value-added service. As important, the news triggers sudden interest among Winnebago's dealers.
- A competitor introduces generous financial incentives for dealers to aggressively push its products, and Potts' customers show strong signs of responding to those incentives.
- Unforeseen promotional bursts from competitors siphon off anticipated sales.
- A competitor suddenly shifts sales and service reps to a market segment that Potts considered secure.
- An enhanced RV quietly and abruptly introduced by a competitor suddenly stirs interest among core dealers.

In all, it would be in Potts' best interest to maintain open channels of communications from a variety of sources to red-flag significant market signals for priority handling. Ongoing intelligence should provide him with some feeling of confidence that he has assembled and screened the data that would produce the best results.

Even in the penetrating light of reality, should he discover that some intelligence is contradictory, false, or contains imperfect knowledge, then he can still lean heavily on his judgment, knowledge of the industry, and years of managerial experience to move forward.

FINDING DECISIVE POINTS

There is the issue of finding decisive points in which to concentrate resources against a market segment. That issue has three dimensions.

First, if Potts is defending a dominant position in any particular segment of the market, which represents a major portion of sales and profits, it is also a choice segment that is likely to be targeted by any competent rival manager who is looking for Winnebago's vulnerabilities.

That means the rival would focus on any product and service weaknesses, pricing practices, communications voids, supply chain gaps, personnel problems, and the overall operating performance of the organization, including its financial stability. That approach would be used for the offense and defense.

Second, if Potts is in an offensive campaign, he also would be looking to penetrate any weakness in the competitor's defenses. Here, too, he and his managers would be using the same type of analysis for developing strategies and tactics.*

Third, beyond the above physical dimensions, there is the morale element that must be considered. The two interact together and should be regarded as inseparable. Morale is fluid, changeable, and subject to a range of circumstances from the quality of internal relationships to the effect a campaign's success or failure has on personnel. Whichever way it moves, the ups and downs of morale spread most easily to affect all levels of the organization.

APPLYING CAUTION

A seemingly iron-clad case has been presented thus far in favor of boldness over caution. Yet, to create such a one-sided argument doesn't cover

---

* For a comprehensive list for selecting defense and offensive areas to pinpoint, see Chapter 1, Table 1.1.

all extenuating conditions that would allow for a more cautious and restrained course of action, such as avoiding a confrontation with a competitor. A decision of that nature could in some instances be considered a bold move in itself.

Caution, in the context of developing and implementing strategy, does not mean timid behavior. Rather, it presupposes that a manager deliberately, and perhaps prudently, chooses to avoid going into a competitive campaign knowing that even if by any reasonable measure he or she achieves a favorable outcome, the confrontation could result in excessive losses of capital, material resources, and staff time.

There are numerous reasons and guidelines that could suggest such a decision to exercise caution. One of the primary guidelines is to test the go-no-go decision by determining if the contemplated campaign coincides with the overall strategic direction and specific objectives of the turnaround plan.

Then, there is a whole range of other considerations that would suggest holding back from entering into a questionable campaign.

Would Winnebago's (or your organization's) staff be mentally and physically able to deal with the pressures of a competitive campaign? That is, are there signs of personnel in a discouraged frame of mind, which would have an adverse effect on morale and, thereby, jeopardize the outcome of a campaign?

This is often difficult to measure accurately. Yet through perceptive observation and by obtaining useable feedback from subordinates, a fair sense of the staff's mindset would surface.

Other considerations include the size and extent of the resources committed by each side, the number of sales personnel assigned to a territory, the availability of after-sales service, the ability to meet delivery schedules, the size of the promotional outlays, and any dedicated offerings or programs that could tip the scale of a competitive campaign.

In another example, one company that had to decide if it would take a cautious approach and stand on the sidelines or enter a competitive conflict is Adesto Technologies, a relatively small company that makes memory chips. It is one of the toughest businesses in the technology industry, and the crowded market is dominated by such big names as Samsung Electronics, Micron Technology, SK Hynix, and Toshiba.

The big four competitors control over 90 percent of the global memory chip business. Given that scenario, a sensible and justifiably cautious approach would have been reasonable for smaller-size Adesto. Instead, the company saw an opening in a market segment overlooked by the industry giants. Adesto went after the growing market in chips for smart gadgets, such as fitness-tracking wristbands and other Web-connected devices, which can't hold enough battery power for the bigger, faster chips made by the larger companies.

Part of Adesto's gamble is that the big names would continue battling over their established turfs by building more powerful chips and avoiding a seemingly insignificant size market. If it works, Adesto will have sidestepped getting enmeshed in a resource-consuming battle that it would not likely win. Whereas, using a niche strategy, it could survive without a direct high-risk confrontation.

Another reason a firm would consider caution is that the type of campaign is not suitable for the firm.* It might be struggling in a market segment that relies heavily on logistics, technical backup service, or providing financial backing that would place one firm at a disadvantage with rivals.

The Adesto example illustrates this point. If it were to enter into a campaign against the industry giants in the powerful chip market that relies on continuing advances in technology, it would turn into a never-ending race with hopes of disrupting the market. And, after an extended period of time, should any of the big firms decide to enter the low-power segment, Adesto could have embedded itself through effective branding, service, and other high-profile approaches to establish a firm position as a strong defender.

Then, there is the issue of a company benefiting from the passing of time. Any number of possibilities exist here, such as waiting for additional funding, making certain that an acquired technology is market-ready, or anticipating the hiring of new key members of a team. While waiting may be justified, it does have a cautionary aspect in that the opportunistic moment may have come and gone.

Another possibility is that after an analysis of the competitor's situation (compiled from information described in Chapter 1, Table 1.1), the rival did not have the wherewithal to sustain a presence in the market. Thus, it would be pointless to move aggressively with a costly campaign. Add to that the possibility that the competitor also was going through a turnaround possibly aimed at an entirely new direction, thereby supporting the decision to avoid a conflict.

There is still another factor, where the opponent appears deceptively stronger than is the actual case, which could be the result of an incorrect

---

* See Chapter 1 for a listing of types of campaigns.

assessment of the rival's real condition. This is altogether feasible, and is often the result of managers exaggerating dangers beyond reality, as some tend to look at the negative side of a situation and come to faulty conclusions. If reevaluated in time, an unnecessary confrontation could be avoided.

These are the some of the principle conditions in which caution would be justified. For the most part, however, success is not achieved by entirely avoiding contact with a rival. That is, if a plan's strategy emphasizes neutralizing the rival's capabilities, an encounter is probable. How, then, can you save yourself from a competitor who is eager to engage you in a combative campaign, inasmuch as you are so eager to avoid it?

As has been discussed in detail, defense is stronger than offense as long as that period is followed by an offensive move. You are only waiting to respond effectively as the aggressor exposes himself and broadcasts his strategy. If, however, you are in a materially weaker position, you can risk going on the offensive against a stronger adversary under the following conditions.

You may be willing to take on some risk and place your bets on a major effort. That condition assumes three possibilities: (1) the corporate culture can sustain such a move; (2) it is not an outright gamble (rather, as in the Adesto case, the market trends were clearly documented); and (3) the leading competitors' strategies were known and Adesto had a clear-cut strategy that focused on moving into a growing market niche that would not be contested for a foreseeable period of time.

MANAGEMENT TOOLS FOR DECISION MAKING

In all of the above situations related to pros and cons of bold action versus cautious restraint, the analysis, decision-making process, and guidelines for implementing the strategies would be best aided by management tools. What follows are guidelines to the classic ones that have been tested and are still in use among many organizations, albeit, in some modified form.

BCG Growth-Share Matrix (Boston Consulting Group analysis)
GE/McKinsey Matrix
Arthur D. Little Matrix
Management by Objectives
Six Sigma

BCG Growth-Share Matrix

With a technique developed by the Boston Consulting Group, this classic model has proved highly useful in assessing a portfolio of businesses or products. BCG Growth Share Matrix (Figure 6.1) graphically shows that some products may enjoy a strong position relative to those of competitors, while other products languish in a weaker position.

Also, each product has its own total strategy depending on its position in the matrix. The various circles represent a product, and, from the positioning of these circles, managers can determine the following information:

- Sales: Represented by the area of the circle
- Market share: Relative to the firm's largest competitor, as shown by horizontal position
- Growth rate: Relative to the market in which the product competes, as shown by the vertical position

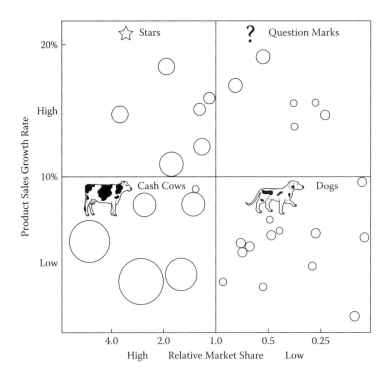

FIGURE 6.1
BCG Growth-Share Matrix.

In addition, the quadrants of the matrix categorize products into four groups:

1. *Stars:* Products that have high market growth and high market share in a fast-growing industry. These products need constant attention to maintain or increase share through active promotion, product improvement, and careful pricing strategies.
2. *Cash cows:* Products that have low market growth and high market share. Such products usually hold market dominance and generate strong cash flow. The object is to retain a strong market presence without large expenditures for promotion and with minimal outlay for R&D. The central idea behind the cash cow is that businesses with a large share of market are more profitable than their smaller-share competitors.
3. *Question marks* (also known as problem children or wildcats): Products with potential for high growth in a fast-moving market, but with low market share. They absorb large amounts of cash (usually from the cash cows) and are expected to reach the status of a star.
4. *Dogs:* Products with low market growth and low market share, reflecting the worst of all situations. A number of alternatives are possible: maintain the product in the line to support the image of being a full-line supplier, eliminate the product from the line, or harvest the product through a slow phasing out.

According to Bruce Henderson, who created the growth-share matrix for BCG in 1970:

> To be successful, a company should have a portfolio of products with different growth rates and different market shares. The portfolio composition is a function of the balance between cash flows. High growth products require cash inputs to grow. Low growth products should generate excess cash. Both kinds are needed simultaneously.

As you review the growth-share matrix, note on the vertical axis that product sales are separated into high and low quadrants. The 10-percent growth line is simply an arbitrary rate of growth and represents a middle level. For your particular industry the number could be 5, 12, or 15 percent.

Similarly, on the horizontal axis, there is a dividing line of relative market share of 1.0 so that positioning your product in the lower left-hand quadrant would indicate high market leadership, and in the lower right-hand quadrant, low market leadership.

The significant interpretations from the matrix are as follows:

- The amount of cash generated increases with relative market share.
- The amount of sales growth requires proportional cash input to finance the added capacity and market development. If market share is maintained, then cash requirements increase only relative to market growth rate.
- From a manager's point of view, cash input is required to keep up with market growth. Increasing market share usually requires cash to support marketing expenditures, lower prices, and other share-building tactics. On the other hand, a decrease in market share may provide cash for use in other product areas.
- In situations where a product moves toward maturity, it is possible to use enough funds to maintain market position and use surplus funds to reinvest in other products that are still growing.

In summary, the BCG Growth-Share Matrix permits you to evaluate where your products and markets are relative to competitors and what investments are needed relative to such basic strategies as building share for your product, holding share, harvesting, and withdrawing from the market.

General Electric Business Screen

The BCG Growth-Share Matrix focuses on cash flow and uses only two variables: growth and market share. The General Electric Business Screen (Figure 6.2) on the other hand, is a more comprehensive, multifactor analysis that provides a graphic display of where an existing product fits competitively in relation to a variety of criteria. It also aids in projecting the chances for a new product's success.

The key points in using the GE Business Screen include:

1. *Industry attractiveness* is shown on the vertical axis of the matrix. It is based on rating such factors as market size, market growth rate, profit margin, competitive intensity, cyclicality, seasonality,

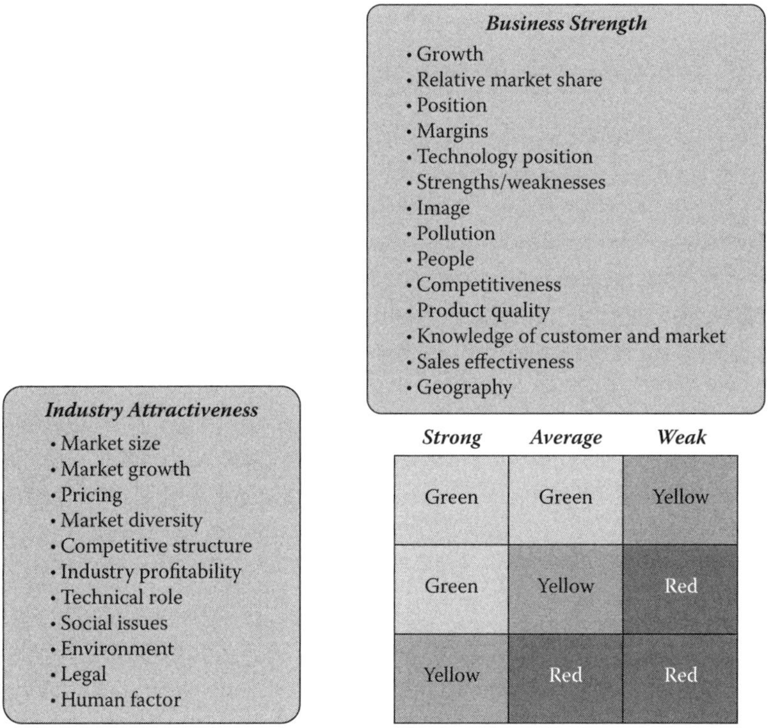

Business Strength
- Growth
- Relative market share
- Position
- Margins
- Technology position
- Strengths/weaknesses
- Image
- Pollution
- People
- Competitiveness
- Product quality
- Knowledge of customer and market
- Sales effectiveness
- Geography

Industry Attractiveness
- Market size
- Market growth
- Pricing
- Market diversity
- Competitive structure
- Industry profitability
- Technical role
- Social issues
- Environment
- Legal
- Human factor

	Strong	Average	Weak
	Green	Green	Yellow
	Green	Yellow	Red
	Yellow	Red	Red

FIGURE 6.2
General Electric Business Screen.

and scale of economies. Each factor is given a weight classifying an industry, market segment, or product as high, medium, or low in overall attractiveness.

Some organizations add or substitute factors that are pertinent to their respective businesses. These include profit margins, impact of a disruptive technology, the law, energy efficiency, environmental impact, entry barriers, and opportunities to outmaneuver competitors.

2. *Business strength* is shown on the horizontal axis. A weighted rating is made for such factors as relative market share, price competitiveness, product quality, knowledge of customer and market, sales effectiveness, and geography. The results show the ability to compete and, in turn, provide insight into developing strategies in campaigns with competitors.

Here, too, organizations customize the list to include management strength, R&D, branding, supply chain management, cost reduction, social media, and customer loyalty.

In actual use, three color sectors divide the matrix: green, yellow, and red. The green sector has three cells at the upper left and indicates those markets that are favorable in industry attractiveness and business strength. These markets indicate a "go" to move in aggressively.

The yellow sector includes the diagonal cells stretching from the lower left to upper right. This sector indicates a medium level in overall attractiveness.

The red sector covers the three cells in the lower right. This sector indicates those markets that are low in overall attractiveness.

Arthur D. Little Matrix

Another time-tested portfolio analysis approach is associated with the consulting organization, Arthur D. Little Inc. In one actual application, a major manufacturer in the healthcare industry used this approach to analyze how its various products stacked up in market share. In Figure 6.3, some of the company's products are used to demonstrate the function of this matrix.

First, note the similarities of this format to the other portfolio analysis approaches already discussed. The competitive positions of various products are plotted on the vertical axis according to such factors as *leading, strong, favorable, tenable, weak,* and *nonviable*. On the horizontal axis, the maturity levels for the products are designated *embryonic, growth, mature,* and *aging*.

The key interpretations for this matrix include:

1. *Nonviable*: The lowest possible level of competitive position.
2. *Weak:* Characterized by unsatisfactory financial performance, but with some opportunity for improvement.
3. *Tenable:* A competitive product position where financial performance is barely satisfactory. These products have a less than average opportunity to improve competitive position.
4. *Favorable:* A competitive position that is better than the survival rate. These products also have a limited range of opportunities for improvement.

Bold Action versus Cautious Restraint • 137

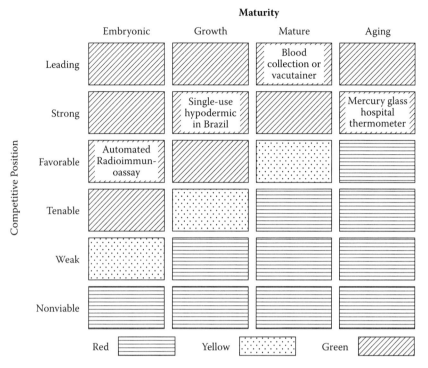

FIGURE 6.3
Arthur D. Little Matrix applied to products.

5. *Strong:* Characterized by an ability to defend market share against competing moves without the sacrifice of acceptable financial performance.
6. *Leading:* Incorporates the widest range of strategic options because of the "competitive distance" between the given products and the competitors' products.

An examination of the four products shows how this matrix worked during a particular period in those products' life cycles.

Automated radioimmunoassay (a diagnostic product used in laboratories) was once considered in its embryonic stage with a favorable competitive position at the time the analysis was prepared for the North American market. This favorable position offered the manager a range of strategy options as long as the decisions related to the overall corporate strategy.

Single-use hypodermic needles and syringes had a strong competitive position in a growth industry. Here, too, strategy options were fairly flexible and depended on competitive moves as well as on how quickly increases in market share were desired.

Blood collection system (vacutainers) had a one-time leading competitive position in a mature industry. To hold existing market share, the company's strategy centered on product differentiation.

Mercury glass hospital thermometers originally had a strong competitive position in a declining industry. This product had less price flexibility. However, by using service, repackaging, and distribution innovations, the company attempted to maintain its strong position before giving in to price reductions. (It, subsequently, has reached the nonviable position and has been replaced by new technologies.)

As in the GE Business Screen, a green-yellow-red system is normally used to indicate strategic options: Green indicates a wide range of options, yellow signals caution for a limited range of options for selected development, and red warns of peril with options narrowed to those of withdrawal, divestiture, and liquidation.

You can use this technique to evaluate your own product and market opportunities.

Management by Objectives (MBO)

MBO, also known as management by results (MBR), is a process of participative goal setting with individual employees. Associated with the management scholar, Peter Drucker, it first became popular in the 1950s and is still widely used. The key feature of MBO is that the manager and employee jointly agree on objectives.

Although implementing the system varies with companies, one successful approach calls for a manager and employee to each develop objectives for that employee in private. At a follow-up meeting, each reviews the other's list of objectives and a negotiated discussion takes place until there is mutual agreement where each signs off on the final list of objectives for a specified period.

To facilitate the process, there is typically a broad category of areas from which to develop objectives, such as personal development, job performance, environmental concerns, cost savings, and various corporate issues. These categories can be customized to suit the manager's agenda for the employee.

An important part of the system is a periodic progress review, with measurements determined by the employee's actual performance against the agreed-on objectives. Thus, one of the major benefits is that employees are totally involved with the goal-setting and with choosing the course of action.

While MBO is associated with individualized one-on-one application, it does have its use with teams by incorporating objectives that are collectively associated with the long-term strategic plan. MBO is perfectly suitable and desirable for turnaround plans, especially when viewed from the varied issues discussed in this and previous chapters.

MBO does have its detractors, however, as in the case where a manager takes a strategic view of where the organization is headed and chooses not to reveal too much information to an individual employee, thereby creating an information gap in the exchange process. Or, a manager chooses not to follow a participative management style and tends to use a more top-down autocratic style and ends up dictating objectives to the subordinate.

Six Sigma

Consisting of techniques and tools for process improvement, Six Sigma was developed by Motorola in the 1980s. Its popularity came about in the 1990s when CEO Jack Welch made it the focal point of his business strategy at General Electric.

The central idea behind Six Sigma is the ability to produce a very high proportion of "six sigma quality" at a defect level below 3.4 defects per million opportunities. However, in practice, this number can vary depending on the process, and an appropriate sigma level can be assigned by management as a way to prioritize areas in need of improvement.

While originally associated with manufacturing, Six Sigma does carry over to such management initiatives as leadership, product design, supply chain management, and cost containment.

One approach to implementing Six Sigma would follow five phases:

1. Define the specific project or management goals
2. Measure key components or decision points of the current process and collect relevant data

3. Analyze the data to determine cause-and-effect relationships and seek out causes of defects
4. Improve the current process based on data analysis
5. Control the future process to ensure that any deviations from the target are corrected before they result in defects

There are numerous spin-offs, variations, and adaptations of the above management tools and techniques in current use by companies. Newer ones continue to arrive, as well, such as General Electric's FastWorks program. Currently in use at GE, it is designed to foster innovation and accelerate product development by first building imperfect early versions, then releasing them to customers and getting feedback, and finally adapting the products where necessary. In application, FastWorks is intended to speed up the rollout of products ranging from lightbulbs to gas turbines to refrigerators.

All programs are intended to produce a competitive advantage in the marketplace. All of them can apply, in some way, to refining your turnaround plan and contributing to developing competitive strategies of the type described in the following chapters.

7
Concentration versus Dispersal Strategy

Chapter Objectives

Be able to

1. compare the benefits of concentration to the dispersal of resources;
2. identify the five categories that comprise concentration;
3. describe guidelines for utilizing a concentration strategy; and
4. indicate circumstances where dispersal of resources is justified.

INTRODUCTION

The essence of concentration is that you are strong in general and in particular at a decisive point. It embodies the theoretical and pragmatic underpinnings that support the entire concept behind the familiar practice of segmenting markets.

Concentration was discussed briefly in Chapter 1 using the example of selecting a segment-by-segment strategy to enter the vast market of China. It was referred to again in Chapter 6 with Winnebago Industries to identify the decisive point for concentration at a defined customer segment of retirees.

Concentration is the primary strategy with which to focus on the vulnerabilities of the competitor. That approach also fits within a framework known as two zones of activity: (1) concentrating on filling the unfilled needs among customers and prospects, and (2) targeting a competitor with the aim of neutralizing those capabilities that prevent you from achieving your goals. Doing so, aligns with the recognizable verbiage of developing a competitive advantage, filling a void, and finding an unserved niche.

How does concentration play out in the marketplace?

Levi Strauss & Co., one of the most recognizable clothing companies in the world, is the creator of the famous Levi jeans. As the acknowledged founder of that category of clothing, it was also part of the fashion explosion of the twenty-first century where denim jeans became a basic item for tens of millions of individuals worldwide. It also turned out to be Levi's problem.

With the continuing trend of vast numbers of individuals wearing jeans, the inevitable parade of competitors followed, such as the likes of Lee and Wrangler, as well as an evolving number of others trying to cash in on the continuing movement. What predictably resulted was intense competition for Levi, with the outcome that its $7.1 billion in 1996 sales nosedived to $4.2 billion by 2003. Over the next decade, sales rose only marginally as the San Francisco-based company attempted a turnaround.

In addition to the marketplace competition, other contributing factors within the Levi organization contributed to the decline. Externally, the company battled a deteriorating retail environment, as both brick-and-mortar traffic and industry-wide denim sales continued to fall.

Internally, Levi's design team was late in latching on to key trends, such as colored denims for women and more tailored jeans for men. Then, fundamental internal issues surfaced, such as a lack of discipline and the inability to correctly identify decisive points on which to concentrate. For several years, some senior-level executives, by their own admission, acted as if they had a monopoly on the denim market and gave little notice when young fashion-conscious individuals began trading in their Levi's for more trendy styles offered by start-up rivals.

Management had lost sight of the rapidly changing marketplace and the new competitors. As one executive pointed out, "We have one of the greatest brands in the world, but I think that there may have been periods where we thought the brand itself could carry us through thick and thin. There's no question that we got complacent."

Industry analysts and former executives confirm that the problems boiled down to complacency and arrogance. Thus, stopping the decline, leveling off, and beginning an upward movement became exceedingly difficult.

Levi's current approach to its turnaround is to grow revenues and profits modestly but steadily, year in and year out. As part of the process, CEO Chip Bergh travelled to the company's worldwide offices and interviewed senior executives about immediate changes for the company and what he could personally do to make a difference. It turned out few senior managers knew of the company's problems and few felt motivated to act otherwise.

After further investigation, it came to light that the embedded company's culture had created a working environment of complacency. Bergh subsequently formed new teams that addressed numerous issues, including initiating programs to control costs, improving internal communications, revitalizing morale, and casting off negative feelings that led to complacency.

He also moved forward by streamlining a cumbersome supply chain, which suffered from an elongated and complicated system of design groups in San Francisco working with Levi's innovation center in Turkey. The result was a repaired system that shortened the entire time period from getting new apparel designs approved to getting the products into the stores.

Bergh and his team then addressed the all-important matter of marketing the iconic Levi brand, which had been woefully neglected. That meant implementing a bold effort by launching a nostalgia-driven campaign highlighting the company's legacy as the inventor of blue jeans.

What are the key concerns in the Levi Strauss case that relate to concentration on decisive points? Numerous issues surface to show the application of concentration within the organization, as well as in the competitive marketplace.

Bergh identified the following issues, which subsequently became part of his turnaround plan.

He faced flat sales and profitability problems through inefficiencies within the organization. The corporate culture was out of synch with a dynamic marketplace, which also impacted innovative thinking. It lagged in recognizing changing consumer tastes, and, once a trend was recognized, the various departments were not coordinated to respond rapidly.

There was also the matter of a lack of discipline within its ranks that reflected on leadership and Bergh's inability to clearly define where to concentrate efforts, which further dissipated efforts. Then, there existed an attitude of arrogance fed by the false notion that the Levi name represented a monopoly.

Worse, too, no effective internal communications system was in place to inform senior-level executives, midlevel managers, and the rank and file about the true state of the company. Outside the organization, Bergh faced a deteriorating retail environment, a growing number of new-wave competitors, and an industry-wide decline in denim sales.

When all of those diverse issues were consolidated, they formed the objectives and strategies of the plan to rebuild the Levi brand. They also served to extend its legacy to former customers, as well as to those Levi failed to reach over the decades.

To implement the turnaround plan, management initially focused on relating stories of the company's origins and its immigrant founder, Levi Strauss, who started the company in 1853. With those stories, it aimed to attract the customers they had and subsequently lost, as well as an identifiable segment of "fans who don't really know who we are." Thus, for Levi, the campaign strategy concentrated on going back to its historical roots as a platform for going forward to grow.

IMPLEMENTING A CONCENTRATION STRATEGY

The essential ideas that you can take away from the Levi Strauss case is that concentration is not a single or even a dual-issue strategy. It consists of a multifactor combination of issues that permeate major areas of an organization. When the issues are looked at from a strategic point of view, they help determine how to concentrate resources at one or more decisive points where you can gain superiority. That includes targeting a competitor's specific weakness or general area of vulnerability.

When these varied issues are grouped, they fall into the following categories:

- Consumers
- Intermediaries
- Competitors
- Regulatory Issues and Industry Trends
- Leadership and Management

Consumers

Know your end-use consumers. Although a seemingly obvious point, it represents a key factor about where to concentrate resources. With Levi Strauss, consumers were neglected, overlooked, or simply taken for granted and subsequently became a central strategy in its turnaround plan.

TABLE 7.1

Bases for Market Segmentation

Demographic Segmentation	Psychographic Segmentation
Sex	Lifestyle and orientation
Age	Psychological and behavioral variables:
Family life cycle	• Personality
Race or ethnic group	• Self-image
Education	• Cultural influences
Income	
Occupation	
Family size	
Religion	
Home ownership	
Geographic Segmentation	**Product Attributes Segmentation**
Country	• Usage/frequency levels
Region	• Product features/benefits
Urban/suburban/rural	
Population density	
City size	
Climate	

In other cases cited earlier, Winnebago had to focus more precisely on its core group of retirees. IBM in its ongoing turnaround efforts emphasizes big data, cloud, and mobile, as it keeps redefining its customer base.

Customers congregate into defined market segments. An advanced grouping of segments was categorized in Chapter 4 as *natural, leading edge, key, linked, central, challenging, difficult,* and *encircled.* Then, there are the more traditional approaches, as shown in Table 7.1 You can modify these criteria to suit your business and industry.

Yet simply labeling categories is not enough. What you are looking for are the behavioral patterns that go beyond the standard segmentation criteria. For instance, what are the primary psychological influences in the buying decision? What personal needs do your products or services satisfy? And, to provide somewhat of a comparative analysis, how well do your products satisfy compared with similar offerings from competitors?

These questions take on a more critical role as the areas of differentiation blur and you are pressed to target decisive points with greater precision. In turn, that calls for the input of valid research—formal studies and informal observation—about noticeable changes in consumers' attitudes, values, and habits.

Intermediaries

Assuming you are a producer of the product or service, your primary interest is in knowing the extent of your dealers', distributors', or retailers' influence in the ultimate path to the end-use customer. Such information includes knowing about the depth of your relationships with each link in the supply chain.

Your object is to gain some measure of assurance that the relationship is mutually beneficial to you and the intermediary. Of course, that knowledge is also of key importance within the framework of knowing how well they serve your needs and how well they contribute to the growth of the end-user market.

Solidifying relationships, in turn, opens a whole range of issues, including compensation systems, special incentives, types of training, financial assistance, and even compatibility. And, then, there is the current emphasis on how far advanced intermediaries are with online marketing and the use of social media.

There are also the pragmatic questions of who is in control within the supply chain as they relate to such factors as:

1. Introducing a new product or entering a new market.
2. Intensifying market coverage where the objective is to increase market share.
3. Making a smooth transition when adding or replacing existing distributors.
4. Changing methods of distribution to complement changes in business strategy, technology, and other movements in the industry.

Introducing a New Product

Only with the appropriate distribution network can you satisfactorily achieve your market-based goals. For instance, as you introduce new products, you may find that your current distributors are ill equipped to provide after-sales services, or they already handle competitive products from other producers.

Or you may be addressing a new kind of clientele and the current intermediaries have no previous experience meeting the delivery schedules and other requirements. Thus, as you enter new geographic markets, the need for appropriate representation may become self-evident.

As you review your share of the business, you also may conclude that your firm is underrepresented, or you may determine that your present distributor network is not going after the business aggressively enough to satisfy you. As a result, you may need to add or replace more distributors in the territory, based on population, sales, buying potential, or other relevant considerations.

By far, the most frequent reason for appointing new distributors is the turnover of existing outlets in the supply chain. These changes may be due to natural attrition, the death or retirement of principals, or the sale or bankruptcy of a distributor. The recent trend toward more specialization or limited-line selling also has led many distributors to drop a particular manufacturer's line.

Often, changes in your distributor mix come about by inadequate distributor performance that leaves the manufacturer dissatisfied. In some instances, there may be an effort to rekindle a past relationship, assuming the distributor is willing to recognize the changing trends in the marketplace and, consequently, makes the required changes in its operating practices.

Intensifying Market Coverage

Rarely should you have to revamp your entire distribution structure to increase market share. However, in some situations, you may want to add intermediaries to distribute your company's products, thus requiring the selection of new distributors.

Table 7.2 highlights the selection criteria most often mentioned by some 200 leading manufacturers in a study on this subject. Look at how the numerous considerations are classified and summarized into a limited number of categories that can apply to any distributor selection. You have the option of modifying or adding criteria to the list to suit your particular needs.

Making a Smooth Transition When Adding or Replacing Distributors

Once you have secured the services of a sought-after distributor candidate and ensured a smooth transition, you must be certain that the relationship brings about maximum benefits to both parties. You need to perform periodic evaluations to stay continually informed about performance.

TABLE 7.2

Criteria for Selecting Distributors

Criteria	Reasons
Financial aspects	A distributor with solid financial strength can assure you of adequate, continuous representation
Sales organization and performance	An exemplary sales record is essential to your potential relationship
Number of salespeople	The more salespeople, the more potential sales and the more effective the market coverage
Sales and technical competence	Salespeople with outstanding technical and sales skills are a marketing advantage and could be a key differentiator against a competitor
Product lines carried:	
• Competitive products	Generally avoid unless circumstances dictate otherwise
• Compatible products	Tend to be beneficial
• Quality level	The higher, the better
• Number of lines	Determine if your line will receive enough attention
Reputation	Your company and product are judged by the image projected
Market coverage:	
• Geographic coverage	Avoid overlaps and conflicts
• Industry coverage	Major user groups must be covered
• Intensity of coverage	Infrequent calls mean lost business
Inventory and warehousing:	
• Technology	Essential to online ordering and maintaining real-time communications
• Kind and size of inventory	Ability to deliver on time, in the right mix, and maintain adequate stock are essential
• Warehousing facilities	Storage and handling must be exacting
Management:	
• Capabilities	Proper leadership spells success
• Continuity	Succession should be assured
• Attitudes	Look for enthusiasm and aggressiveness

These evaluations may be in the nature of specific operating appraisals or may take the form of overall performance reviews. If they are simple and limited in scope, you could conduct them monthly. Thorough analyses, however, should be undertaken only at infrequent intervals: annually, biannually, or even triennially.

If you engage in selective rather than exclusive distribution, the amount of evaluative input that you can obtain from your distributors is quite limited, forcing you to rely mostly on your own records, observations, and intelligence. If your product is a high-volume, low-cost item with little need for after-sale servicing, you can restrict yourself to a more limited evaluation than in the case of complex systems installations.

If your team is composed of many hundreds of multiline distributors, you will tend to take a closer look at a particular reseller only if its sales trends are way out of line. This procedure is called *evaluation by exception*.

If, in contrast, your firm employs only a moderate number of outlets, your analysis can be more thorough. You may not even need a formal evaluation if you have a close, continuous working relationship.

Changing Methods of Distribution to Complement Changes in Business Strategy and Movements in the Industry

When making changes in business strategy, there are key functions you have to deal with in shaping an effective supply chain. The following listing highlights the dominant factors to consider:

- *Information*
    - Collect, analyze, and disseminate market intelligence about potential and current customers, competitors, and other forces affecting the market.
- *Communication*
    - Combine various forms of communication including literature, videos, and workshops to attract and retain customers.
- *Negotiation*
    - Seek agreement on price, terms of delivery, and other value-added services as they relate to a preferred-customer status and long-term relationships.
- *Ordering*
    - Set up an easy-to-use system for the efficient transmission of initial orders and for use with automatic resupply requirements (where applicable).
- *Financing*
    - Develop the means to fund improved electronic control system.
- *Risk Taking*
    - Assume the responsibility for risks associated with the expanded role and activities of middleman.
- *Physical possession*
    - Develop a suitable capability to store additional varieties of products for customers and manage increases in inventory turnover.

- *Payment*
  - Design an effective system for payment, including the selective financing of inventories for the buyer.
- *Title*
  - Develop a system to pinpoint the transfer of ownership from seller to buyer. In some situations, inventory is held at the buyer's location and title changes only when usage occurs.

With the backward and forward flow of activities throughout the distribution channel, different participants in the channel assume distinct functions. Therefore, whether manufacturer or distributor, when forming relationships be certain that roles are clearly defined for each channel member.

Competitors

Competitor guidelines were provided in Chapter 6. In addition to determining their strengths and weaknesses, you want to know such information as: What is their product mix and are there any gaps that would create an opportunity for concentration? Also, is any particular competitor's participation in the market growing or declining? What are the reasons in either situation? Which competitors may be leaving the field? Why? What are the implications for us?

Then, there are more penetrating questions about new domestic competitors that may be entering the market. What market niches are they filling? What comparative advantage are they using?

Are there opportunities for cooperative joint selling, as in the case of IBM and Apple coming to an agreement to partner on developing services optimized for iOS, the iPhone maker's mobile operating system? The two companies' sales teams had been co-training to make sure they were well-equipped to cross-sell each other's products.

As for new foreign competitors, are there any serious contenders on the horizon? How threatening are they? Which competitive strategies and tactics appear particularly successful or unsuccessful when used by competitors?

Finally, what new directions, if any, are competitors pursuing? Are there new strategies as they relate to markets, products, pricing, marketing, supply chain, disruptive technologies, leadership, or others?

Regulatory Issues and Industry Trends

An ever-present cloud of regulations from global to local levels places constraints on business operations. For instance, the most visible companies affected by restrictive government regulations include such high-profile companies as Apple, Volkswagen, Fiat Chrysler, Microsoft, and Qualcomm; each is trying to fight an uphill battle to expand their respective operations.

Then, there are strong Internet data protection barriers faced by Google and Facebook in Europe and elsewhere. Collectively, the alleged charges range from threats to national security, to price fixing, and monopolistic behavior.

In the United States, various national and local groups provide regulations that must be addressed. For example, the Occupational Safety and Health Administration (OSHA) now requires companies to notify the agency within 24 hours about anyone who gets admitted to the hospital with an injury sustained at work.

Thus, the key issues deal with what are the immediate and long-term consequences of the regulations as they affect a turnaround. What needs to be done to comply with the rules? And, of course, what political and legal developments are looming that will improve or worsen your situation?

As for industry trends, what threats or opportunities do advances in technology hold for your company? This question was a priority one in 2014 for IBM as it labored through a turnaround after experiencing a period of nine quarters in a row of shrinking revenues.

The giant company also was undergoing a downturn in its traditional IT businesses. Some of IBM's lower-margin businesses were sold off, staff cuts were made, and other defensive cost-cutting measures put in place. However, as was pointed out in detail in Chapter 5 (Defense versus Offense), defensive moves must be followed by offensive moves, which is exactly the pattern followed at IBM.

That is where CEO Virginia "Ginni" Rometty went on the offensive by recognizing the escalating trend in cloud services and, subsequently, acquired cloud services provider Soft-Layer Technologies. Noticing other specialized technology-related movements on the horizon, she boldly invested heavily in developing and commercializing Watson, a reasoning-computing system capable of sifting through millions of scientific papers in seconds.

Beyond technology, what are the trends concerning the environment, cultural and geographic shifts, political movements, and the "green" movement that impact a concentration strategy? Then, there are the pragmatic trends concerning sources of financing, replacing equipment, accessing raw materials, and locating new suppliers.

Leadership and Management

Leadership was discussed in detail in previous chapters. Chapter 2 dealt with interpersonal relationships and a range of expectations from staff. Chapter 4 focused on guidelines related to the personal qualities of courage, intellect, intuition, and determination.

What follows are three issues that concern the managerial aspects of leadership, especially as they relate generally to a turnaround plan and specifically to a concentration strategy: *market research, planning,* and *the organization.*

Market Research

Market research and, particularly, competitive intelligence are the foundation underpinnings for developing strategies. Acquiring quality research and using it wisely improves your ability to implement concentration strategies with greater precision. Accordingly, it is advisable to prioritize that function by establishing (or expanding) an internal market research function that provides ongoing intelligence, rather than rely only on outside sources on an as-need basis.

To augment the process, it is best to involve personnel at all levels, beginning with salespeople who interface with the customer. They are in an advantageous position to observe competitors' moves real-time.

What follows is a workable feedback system where assigned individuals assemble information and disseminate key findings to those individuals responsible for implementing the plans. That also means training the research people to systematically sort through data to pinpoint a competitor's strengths and weaknesses, as well as to locate decisive points through which to concentrate resources (see Chapter 6, Table 6.1).

Based on the significance of the research, you can then determine with greater accuracy how to position a new product or service against a competitor's position before launch. Further, where market research is effective, it helps in testing a new product's or service's acceptance among customers before a full-scale introduction campaign gets underway.

There is still one additional piece that goes into an effective research system. It is the utilization of a special category of individuals who would gather intelligence: agents. Within legal and ethical bounds, they can be valuable additions to your efforts. A description of five categories of agents and their uses are found at the end of this chapter.

*Planning**

As a leader, you would act as the key driver in the planning process. In which case, you could add immeasurably to the quality of the plan by using a participative approach that seeks the collaborative input from individuals in other functional groups within the organization. The outcome, more times than not, leads to higher levels of morale and results in employee buy-in at the critical stages of implementing the plan.†

Whether planning takes the ultimate form of a turnaround plan, a strategic business plan, or a business unit's tactical plan, it should be a formalized procedure within your organization that results in a series of actions or campaigns, and it shouldn't end up as the proverbial leather binder on the shelf.

As the planning process proceeds, the following checklist would be useful to monitor its progress:

- Do your long- and short-term objectives complement the strategic direction of the plan?
- Do you prioritize the objectives and make certain there is acceptance and participation among various functional groups with the object of reducing the chances of friction?
- Are your objectives realistic, achievable, and measurable?
- Do you utilize appropriate metrics to assess performance and make essential midcourse corrections?
- Are you budgeting sufficient resources to achieve your objective?
- Does each objective have a corresponding strategy?

---

* The term planning is deliberately used here rather than *plan*. As former U.S. president and general, Dwight D. Eisenhower adroitly stated, "The plan in nothing; planning is everything." His concern and interest was in the thinking, deliberating, and strategizing that stemmed from a collaborative approach and resulted in a document called a plan.
† See Chapter 2, Table 2.1.

- In the analysis of market segments, is there a procedure to locate decisive points in customers' needs and in competitors' defenses through which to concentrate?
- Is there a systematic screening process to identify opportunities and threats, such as a SWOT (strengths, weaknesses, opportunities, and threats) analysis[*] or similar approach?
- How effective are you in integrating what-if scenarios into your plan and, thereby, preparing to react rapidly to competitive challenges and threats to your business?

An overriding consideration throughout the planning process is that the strategic direction, objectives, and strategies are aligned with your corporate culture—and with the future direction of the market. This point is especially significant in developing your turnaround plan.

Referring again to IBM's Virginia Rometty, she best expressed this idea as, "If I have learned nothing else in all my years here, my biggest lesson is you have to constantly reinvent this company. That's how you get to be 103 years old." That viewpoint, in part, is what planning a turnaround is all about.

Organization

A central issue about managing within the organization is that the strategic and turnaround plans coordinate with the other business units' and product lines' plans. Consequently, keeping in mind that a top-to-bottom integration leads to a united effort, several points need attention.

First, monitor the plan to determine its relevancy in a changing market where consumer behavior tends to be changeable, intermediaries often are in a positioning fight for power, competitors are in various stages of offensive and defensive actions, regulatory issues and industry trends keep evolving, and leadership and management are under continuing scrutiny.

Second, determine if an effective communications network is functioning within the organization and if it contributes to forming a cohesive spirit among personnel.[†]

---

[*] The process for conducting a SWOT analysis is found at the end of this chapter.

[†] The average *Fortune* 500 communications department employs 24, though the size can range from 3 to 600 depending on how extensive its roles, including marketing, investor relations, and government affairs.

The following model is one such communications system that can permeate the organization and provide connectivity.

It begins with an e-communications platform that permits executives to actively connect with their people through personalized video messages combined with professionally prepared informational content. The notion is founded on the rock-solid principle that competent, trained, motivated, and well-informed employees are the essential elements in turning around a strained organization.

To those ends, the system is built on the following five realities that define today's competitive marketplace:

1. Implementing even the most basic competitive strategies requires a well-developed corporate vision, strategic direction, and motivated employees.
2. Foundation business and strategy principles do exist and need to be followed regardless of the type of business.
3. Effective decision making depends on nurturing a customer-driven company culture consisting of core values that must be communicated consistently to expect winning performance.
4. Business excellence relies on persistently searching for best practices, then benchmarking them selectively for application to the firm.
5. Looking for fresh opportunities requires managers to cultivate in their employees a wide-vision lens. As geographic distances dissolve, cultural differences emerge as the more important possibilities for growth.

The president of a large entertainment network declared:

Managers must keep faith with their long-term vision, but now they need to set and motivate staff for quarterly and annual targets. They need to inspire whole groups of people to cushion the company from the growing pressures from outside.

Guidelines to Utilizing a Concentration Strategy

The following summarizes the various influences and applications of a concentration strategy, particularly the ones relating to planning within the organization:

First, recognize that concentration is a strategy with which to challenge larger competitors by using a segment-by-segment approach. The outright advantage is that it permits you to concentrate your resources to gain a superior position at a decisive point, even though it might create some exposure elsewhere.

Second, market and competitive intelligence are essential to target a segment for initial market entry and to expand toward additional growth segments. Pinpointing segments also includes employing advanced segmentation techniques (Chapter 4) and the traditional techniques described above. Another approach is to identify emerging, neglected, or underserved segments.

Third, part of the process of targeting segments of opportunity requires a level of discipline and objectivity in knowing when to exit underperforming ones. Then, you are better able to concentrate your resources and serve a dedicated group of customers.

Fourth, in keeping with the classic approach of niche marketing, you can design products and services to meet the needs and solve the problems of earmarked customers. This is the primary reason for coordinating the strategic and turnaround plans with the other functional plans of your organization.

Fifth, it is a key point of concentration to understand that every market segment presents opportunities to fill market gaps, allocate resources efficiently, and exploit a rival's limitations. To those points, you want your organization to become practiced in seeking out and uncovering gaps of opportunities.

DISPERSAL STRATEGY

Concentration is in sharp contrast to the overly common planning approach of dispersing resources in several directions, covering numerous objectives, segments, and isolated actions. Whereas the thinking may be to play it safe and cover all contingencies, instead it has the potentially damaging effect of dramatically exposing weaknesses and revealing areas of vulnerability. The result is that the chances of failure multiply through the excessive thinning-out of resources.

Therefore, as a general rule, no resources should be allocated unless the need is specific and urgent. That is, unless a valid opportunity opens up and it is in sync with the overall plan, or there is a veiled competitive threat that could materialize into a serious problem.

For instance, there may be situations where no market segments have been singled out that offer long-term growth, yet there are strategic reasons to maintain some presence. Even where segments can be selected, there are no dedicated products or service to offer as a differentiated product benefit, nor is there any reasonable chance of achieving a noticeable competitive advantage.

Notwithstanding, the fact remains that resources have been divided and separated countless times, without the leader having any clear reason for it, simply because he or she vaguely felt that this was the way things ought to be done. This viewpoint should be avoided as soon as concentration is recognized as the preferred strategy, and every dispersal of effort should be considered an exception that has to be justified.

That means determining how you and your competitor's resources are deployed by segment, which could be viewed as exceptional secondary opportunities that look particularly rewarding. To repeat, however, there has to be a clear indication of an identifiable decisive point in which to concentrate resources to justify diverting personnel, capital, and other resources and risking needless exposure in the primary segment.

Consequently, a primary task in planning a campaign is to identify the rival's decisive point, the second task is to ensure that the forces used against that point are concentrated. A third task is to keep each minor campaign as subordinate as possible and make sure it doesn't draw off too much strength from the main objective.

Seen in this light, it is advisable to operate offensively only in the main area and to stay on the defensive elsewhere. Going on the offensive would only be justified if outstanding conditions should occur.

Moreover the defensive mode at the secondary points should be maintained with the minimum of strength, if possible, unless there are exceptions, such as joint venture possibilities, similar to the previously cited combined marketing effort between IBM and Apple.

Still other exceptions may be justified, as in the instance where high-profile competitors battle for dominance in specialized segments. Such is the case with the intense competition going on in the race to dominate the tablet market.

Apple, Nokia, and Microsoft each introduced new tablets on the same day in October 2013. Those devices competed for consumer attention against several models released around the same period by other tech heavyweights, including Amazon and Samsung. As the proliferation of various types of tablets accelerated and the skirmishes intensified, the market became more fractured and the competing companies moved to a niche strategy.

Microsoft, for example, focused primarily on professionals by offering tablets that double as PCs. Samsung went for a large variety of tablets, some that included a stylus for drawing and taking notes, to cater to different professions and interests. Apple marketed its iPads as versatile devises for both work and play. Amazon offered low-priced tablets to get people to buy content from its stores.

These instances are market realities, which make concentration strategy and most other forms of strategy more art than science. Yet, with any movement to concentrate resources on a segment of the market, there are generally accepted criteria you can use for offensive and defensive campaigns.

It is a major act of strategic judgment to distinguish among possible decisive points in the competitor's markets in which it operates. Your task, then, is to estimate the effects of concentration versus dispersal and their possible effects on the rest of the plan.

In a major campaign, the more means you can concentrate, the more certain the effect will be. Consequently, any partial use of resources directed toward an objective that does not bring about success should be avoided.

These gradations of concentration and dispersal vary depending on the individual case. For example, McDonald's, the fast-food chain, while carrying on its main-line business, became aware that individuals between the ages of 18 and 32 represented but one segment, albeit an enormous group, in which they needed to retain a strong foothold for future growth.

The company diverted resources to focus on that demographic segment known as millennials. As for competition, several feisty adversaries, such as Five Guys, Chipotle, and Subway, also were aggressively going after that group, which posed a question of how much resources should go into the effort.

While carefully monitoring competitive moves, at the same time observing the group's changing tastes, McDonald's decided to launch its market response with a new food offering: McWrap. The product was a chicken-based tortilla that could be made to order in 60 seconds and would qualify for McDonald's stringent fast-food requirements of time, speed, and convenience.

Continuing with its targeted strategy, McDonald's also recognized the diversity of tastes in global markets where there were numerous decisive points in which to concentrate. The company permitted alterations in its basic recipe to meet indigenous tastes, while at the same time allocating just enough resources to defend against the offerings of local competitors.

The McDonald's case is intended to demonstrate the general reasons for dispersing resources. Management knew there were two conflicting interests involved: (1) there was an opportunity to gain a firm hold in a specific segment that represented a growth opportunity and would justify siphoning off product development and marketing resources to go after millennials; and (2) management understood the ramifications of moving too far afield from concentrating at decisive points.

In summary, finding the decisive point that focuses on keeping your forces concentrated is an essential task that should occupy your thinking as you shape a strategy for your company, business unit, or individual product. Choose correctly and you are likely to win the competitive battle. Choose incorrectly and you could end up spreading your resources thin and battling it out in the marketplace, which ends up consuming excessive amounts of human, material, and financial resources.

UTILIZING AGENTS FOR COMPETITIVE INTELLIGENCE

Agents explore the human side of competitive intelligence by reporting on the behaviors and personalities of key individuals. Their primary means of assembling information is through personal interaction and observation. Agents also are effective in screening and interpreting events, news, and authenticating information that come in through other sources.

As indicated earlier, be certain that you are not violating any legal and ethical guidelines set forth by your company or industry groups. Also, before selecting prospective agents, attempt to find out their special skills and, most importantly, why they want to undertake such a task. Once you know their motivation, you can determine how best to employ them.

For instance, some individuals' only interest is money, with minor interest in obtaining accurate information about the competitor's true situation. In such cases, question their integrity and use great care in using them. Also, be sure to communicate clearly the specific information you need.

The following represents categories of agents, along with suggestions on how to use them.

General Agents

General agents is an overall category that includes individuals with whom you would normally interact during professional gatherings. They tend to freely share company information to satisfy their personal interests, such as making new industry contacts and advancing their careers.

Often, they are somewhat uncaring about their respective company's security, or they are simply oblivious to the dangers of revealing company secrets.

General agents are found in a variety of places. Trade shows are excellent places for gathering intelligence from these individuals. They also are in places where competitors typically reveal extensive information through elaborate demonstrations about their products and where they liberally distribute literature overflowing with facts about pricing, backup services, logistics, product specifications, and so on.

Also, key individuals from competitors' organizations often present technical papers at open meetings that detail sensitive information about upcoming products, services, and even market-entry plans. Then there is the Q&A period that usually follows where the speaker, trying further to impress an audience, inadvertently pours out more sensitive information.

Another prime area for intelligence gathering is the familiar hospitality suite at trade shows and professional meetings where alcohol and talk flow easily. It is a spot where security is often lax and everyone's guard is down.

College classrooms represent another source of information, particularly where adjunct professors and instructors are also executives and use their respective companies as case examples. In some instances, where fellow students work in companies that interest you, useful data are often revealed about their companies through class presentations, term papers, and casual conversations.

Inside Agents

As with general agents, inside agents work for competitors. In many cases, they may have been bypassed for promotion, feel underpaid and underappreciated, relegated to insignificant jobs, or generally pushed aside in a variety of political or power struggles within the organization.

They feel abused and see their careers languishing unless they make some bold move. They also may find themselves surrendering to financial pressures to keep family and self whole, and their attitude may be now-or-never.

You need to assess such individuals carefully for their stability and determine how to use them judiciously. Obviously, you want their information, within the bounds of ethical and legal guidelines.

Beyond personal observations, you would employ inside agents for their expertise to sort out meaningful information from scientific and professional journals, industry studies, or from innovative projects described in articles and professional papers written by the competitors' employees.

Product literature and product specification sheets readily available at trade shows and meetings are packed with tremendous detail. Your agents should be able to interpret the data for meaningful intelligence.

In-house company newsletters and news releases contain a fountain of information about individuals who left a competitor's employment and may have moved to the consulting circuit. If approached, these former employees may be willing to reveal information, unless specific contractual restrictions apply.

Press releases may include new employee announcements along with job descriptions, contracts and awards received, training programs available, office or factory openings or closures, as well as specific news that reveal competitor's activities. Here, again, your inside agent could handily provide useful interpretations.

Beyond the above listing, there is the continuing flow of rumors from customers and suppliers that your agent can sort out and verify. Additionally, there are local sources worth tapping, such as banks, local trucking companies, and real estate offices.

Double Agents

These agents try to extract intelligence about your company. Stay alert to their intentions. Once identified, you could attempt to turn them around and get them working on your behalf.

They would then serve in the same capacity as inside agents. Here, too, you can assume double agents seek lavish rewards and may even show similar personality traits and motivations as inside agents.

However, it is in your best interest to exercise caution. That is, determine the veracity of these individuals, the reliability of their information, and how long you can expect them to remain loyal to your cause. Once again, make certain you are not violating ethical, legal, or policy guidelines.

Expendable Agents

These agents are your own people who are deliberately fed inaccurate information, which is disseminated in a variety of ways to cause competitors to make wrong decisions. These contrived leaks take many forms. For example, passing fabricated information about new product features through sales reps who come in contact with competitors' reps, or product managers revealing false dates about a product launch that would disrupt a competitor's plans.

In spite of your possible discomfort when undertaking such activities, look at the situation from strictly a strategist's viewpoint. Misinformation needs distribution to divert competitors from directly opposing your strategy moves.

You, thereby, preserve your company's hard-won market position, control needless expenditures of financial and human resources fighting unnecessary market battles, and avoid disrupting your strategies.

Living Agents

These agents usually provide the most credible information. They are generally experienced, talented, and loyal individuals who can gain access to, and become intimate with, a competitor's high-level executives. They sit in a position to learn their plans and observe movements. These individuals are truly the eyes and ears and often enjoy the closest and most confidential relationships.

Perhaps the one unsettling issue to cope with when using agents—but certainly worth knowing—is which of your employees intentionally or inadvertently passes on your company's information directly or indirectly to competitors. Eventually, these individuals are exposed and you can obtain valuable clues about what motivated them to those acts.

Another concern is that engaging in such stealth activities is usually contrary to the type of practices most managers care to undertake. Again, think of business intelligence as essential to running a company in a highly competitive environment.

Above all, it is indispensable to the development and integrity of competitive strategies. To that end, Table 7.3 provides general questions for

TABLE 7.3

Questions You Should Ask of Agents

- What are the competitor's overall business strategies, particularly ones that would impact my plans?
- What is the competitor's financial picture, including breakdowns of costs and sales by product lines?
- What new products or services are under development?
- What new markets are expected to be targeted?
- How are the competitor's business units staffed and organized, especially in marketing and product development?
- What is the caliber of the competitor's leadership?
- What is the caliber of employees?
- What market positions or market shares do key competitors hold within each product segment? Are there plans to increase, maintain, or reduce their respective positions?
- Where are the competitor's vulnerabilities, which could represent a decisive point for a concentration, e.g., product depth, product quality, customer service, price, distribution, marketing, and reputation?
- What is the culture of the organization, e.g., aggressive, passive, complacent? Is there any behavioral effect on the leadership and employees?

guiding you in making the best use of agents. You may wish to add your own specific questions that are especially pertinent to your business and competitive situation.

HOW TO CONDUCT A SWOT ANALYSIS

How you choose to prioritize your actions depends on how you assess your company's or group's strengths, weaknesses, opportunities, and threats. One familiar diagnostic tool to estimate the situation and determine how and where to concentrate your strength is the SWOT analysis (Table 7.4). It is a widely used and time-tested approach, especially within the framework of a concentration strategy.

When employed in a group setting, it provides a highly reliable technique for estimating your situation from internal and external vantage points. You also have options of adding complexity to the analysis by using a quantitative weighting system to grade each of the items you wish to evaluate. Or, you can simplify the use of SWOT by referring to the points as a guide in an informal, freewheeling discussion.

TABLE 7.4

SWOT Analysis: Strengths, Weaknesses, Opportunities, Threats

Strengths	• Look objectively at your organization's or group's unique strengths, not just its physical resources. • Identify those special skills inherent in your organization that would permit you to push the boundaries of innovation and discovery. • Single out any unique characteristics in corporate culture, leadership, internal communications, products, systems, technologies, and processes. *Now, do the same analysis about your competitor.*
Weaknesses	• Determine what weaknesses you see among the above factors. • Look at possible choke points that could prevent implementing business plans. • Examine what can be revamped, reorganized, or discarded. • Estimate at what costs in time, money, and resources weak points can be remedied. *What weaknesses can you detect about your competitor?*
Opportunities	• In viewing the customer, competitor, industry, and environmental situations, what opportunities do you see that would represent a decisive point? • What openings exist to displace the competition, expand the company's entry into new markets, serve new customer groups, or generate new revenue sources? • How would you define the opportunities for long-term growth versus short-term limited payouts? *What similar opportunities would your competitor have against you?*
Threats	• What immediate and longer-term threats do you anticipate and how are you going to face them? • Are advances in technology outpacing your company's financial and human capabilities to keep up? • What governmental, environmental, or legislative issues are looming to hinder your ability to operate profitably? *What point of concentration can competitors use to threaten your market position?*

Even with your best efforts at conducting an accurate SWOT analysis, the reality exists that your turnaround plan comes apart when your original estimates about market conditions don't materialize.

Unforeseen situations can loom as potential threats, such as unexpected price wars, disgruntled channel members along the supply chain, changing industry priorities, or shifting demographics in your primary segments. Then, there are threats from overly aggressive competitors that can hamper your best efforts.

Therefore, build enough flexibility and what-if scenarios into your plan. Also, develop second-tier objectives in the event the primary ones are no longer within reach. Then you can react in sufficient time to remedy situations that might otherwise deteriorate beyond a reasonable chance of recovery.

8
Indirect versus Direct Strategy

Chapter Objectives

Be able to

1. identify the three underlying principles governing an indirect strategy;
2. describe how the physical and psychological components of an indirect strategy interconnect; and
3. incorporate indirect strategies into a turnaround plan to circumvent strong points of competitive resistance.

INTRODUCTION

The roots of indirect strategy can be traced to antiquity. The basic concepts derive from military strategy, which is also the source of most every major strategy principle practiced through the centuries, as well as those modern-day applications described in these chapters.

The following quotes cite the origins and references to indirect and direct approaches:

> There are not more than two methods of attack: the direct and the indirect. Yet these two in combination give rise to an endless series of maneuvers.
>
> **Sun Tzu, ancient Chinese military general, strategist, and philosopher (c. 544 BCE–496 BCE)**

> History shows that, rather than resign himself to a direct approach, a Great Captain will take even the most hazardous indirect approach. He prefers

to face any unfavorable condition rather than accept the risk of frustration inherent in a direct approach.

Sir B. H. Liddell Hart, a British soldier and renowned military historian (1895–1970)

The concept of war (conflict) does not originate with the attack, because the ultimate object of attack is not fighting; rather, it is possession.

Carl von Clausewitz, a Prussian general and military theorist (1780–1831)

The direct and the indirect lead on to each other in turn. It is like moving in a circle; you never come to an end. He who knows the art of the direct and the indirect approach will be victorious. Such is the art of maneuvering.

Sun Tzu (c. sixth century BCE)

To a great extent, an indirect strategy is an extension of the previous chapter's discussion on concentrating at a decisive point. One of the primary principles underlying the indirect approach is that, where possible, you avoid a costly direct confrontation with a competitor. Rather, your strategy would circumvent your rival's strong points of resistance by means of maneuver. At the same time, your aim is to serve the current and evolving needs of your markets.

The following case example introduces you to the diverse applications of indirect strategy.

Uber, the car-hire company, utilizes an indirect strategy as part of its growth strategy: (1) to rapidly expand into additional segments of domestic and global markets; and (2) to attack competing car-hire companies by attracting their drivers as it attempts to circumvent obstacles from regional taxi commissioners and government officials. One of Uber's maneuvering strategies is to assist its drivers to finance car purchases, especially among those individuals who have poor or no credit history.

To implement the strategy in the United States, Uber partnered with General Motors, Toyota, local car dealerships, and several financial institutions to reduce drivers' monthly payments and get them on the road faster with a new car, with the positive effect of improving the quality of service among passengers who enjoyed riding in the late-model cars. (Uber doesn't permit its drivers to use cars older than 10 years.)

The San Francisco-based company views its middleman role as an indirect strategy to rapidly grow its network in cities worldwide. Uber does no lending; it enables drivers to borrow at better rates than they would otherwise get on their own. To add a measure of security and legitimacy to the program, just as Uber deducts its commissions from customer payments on its drivers' behalf, it also makes automatic loan payments, remitting them directly to the auto lenders.

Uber's indirect strategy was spearheaded by its car financing program. Yet it impacted such areas as competition, regulatory obstacles, driver acquisition and retention, payment systems, relationships with vendors and related services, which are all potential drivers of market expansion.

The above issues compress into three underlying principles that feed the process of developing an indirect strategy. They include:

1. Think strategically
2. Maneuver tactically
3. Unbalance the competitor

Think Strategically

Uber's indirect strategy fit perfectly with its strategic thinking about going global into overseas locations, such as Paris, London, and scores of cities from Manila to Milan, as well as its expansion into New York, Boston, Chicago, and Washington, D.C. Whereas an indirect strategy outwardly may appear as a local-based promotion, its core foundation actually starts as a platform for broader applications. That said, the underlying concept should be long-term, strategic, and global, yet flexible enough to be tweaked to satisfy local conditions.

In another instance, Cisco Systems prospered in its core business by selling networking equipment to almost every business with a website. However, with the introduction of cloud technology, many of its customers stopped buying that equipment and now rent computing capacity from such companies as IBM, Microsoft, and Amazon. Thus, the need for a turnaround plan.

The planning process led Cisco to redefine its direction as "the great connector," which is defined by a system they call *Intercloud*. The system consists of a set of software tools that businesses can use to easily shift their IT workload among their own data centers and various cloud services.

To secure their new position, Cisco also set up an online marketplace for Intercloud-compatible software designed by other vendors that would provide specialized applications as an inventory management tool for auto dealers or a service permitting retailers to rent additional computing power.

With the new direction, Cisco is indirectly circumventing any direct confrontation it would face against its longtime customers by avoiding entering the cloud business. Meanwhile, Cisco is acquiring technology and persuading cloud providers to join the Intercloud ecosystem and thereby remain an integral part of the industry movement.

What does thinking strategically mean? First, it is akin to strategic planning and its specific application into turnaround planning, in that the output of ideas, concepts, and innovations help to shape the plan's objectives and action strategies. Whereas strategic planning may be a formal process that takes place periodically, strategic thinking is an ongoing mental activity in the quest, for instance, of finding the next frontier for your company or group.

Second, thinking strategically spans time and space. It transcends the borders of nations and markets and steers the mind to think of innovative strategies and tactics. As expressed in the above company examples, it is the actualization of the familiar phrase: Think globally and act locally.

Within that space, there also are the pragmatic considerations when thinking beyond your natural borders. Therefore, your thoughts should focus on how to outthink, outmaneuver, and outperform existing and emerging competitors.

What is the impetus that compels you to think strategically and envision the future? Consider the following issues and then add your own industry, company, and competitive issues to the list:

- Intensifying competition from developing countries shocked many traditional-minded executives into devising fresh strategies to respond to competitive prices that often ranged from 30 to 40 percent below prevailing market pricing. Now, shifts in energy sources, human rights concerns, and higher wages in those countries that originally offered low-cost manufacturing have forced fresh strategic thinking about the future.
- Changing market behavior driven by new mobile devices and the continuing shift to Internet buying, along with new flexible manufacturing techniques, convinced even the most skeptical executives about vast new opportunities and competitive advantages of reaching and distributing specialized products and services targeted at dissimilar customer groups.

- Shifting behavioral lifestyles influenced marketers to focus on how different groups live, spend, and act—all of which were being highlighted by the media and influenced by diverse political, economic, cultural, and social movements.
- Shortening product life cycles due to the proliferation of new products and the continuing flow of dazzling new and affordable technologies convinced executives to probe for emerging or previously unserved market segments. In turn, those circumstances triggered even greater efforts to push for faster, cheaper, smaller, better products.
- Continuing pressures on profitability and productivity activated the pervasive movement toward downsizing, reengineering, and outsourcing. The result: A rush by many forward-looking executives to create market-sensitive organizations committed to total customer satisfaction. With that movement came a demand for attracting the right type of people to managerial positions with the training and experience to take on the essential responsibilities.
- Disruptive technologies, skyrocketing progress in Internet commerce, cybersecurity, widespread industry regulations and deregulations, new sources of energy, need for substantial financial resources, and the expansion of cross-ocean relationships in Asia and Africa created additional challenges for executives to think, anticipate, and act.

Strategic thinking, then, encompasses numerous issues. Yet, to avoid being totally overwhelmed by their complexity, consider the following six-part process to activate your mind to think strategically, but with goal-driven purposes:

1. *Use a far-reaching lens.* Focus on what are the leading-edge trends that would affect your core markets, and include peripheral areas that represent growth opportunities, as well as expose vulnerable areas through which competitors can attack you. As part of the process, build outside networks of individuals and information sources that can expand your vision and thinking, beyond your current boundaries.

   Think about the advances in technology made possible by joint projects, such as Johnson & Johnson using Watson, IBM's big-data service, to assess and evaluate medications before producing them for the mass market. What used to take J&J scientists decades to analyze vast amounts of genomic data and patient histories, now is handled in just a few days. The analysis further yields the benefit of pursuing innovation as a basis for differentiating J&J products.

2. *Think critically; question generally accepted dogma.* The sources of information through such outlets as seminars, workshops, and the media attempt to keep you current by communicating information, as well as the latest catchwords and terminologies. It is for you to question and transpose the data, news, reports, concepts, and underlying assumptions to seek additional meanings that would open your mind to new pathways of thought.

3. *Interpret and extrapolate.* As an extension of the above, synthesize information from a variety of sources before developing an opinion. Look for convincing patterns among those multiple sources of data. Then verify your conclusions by matching them against the strategic direction and objectives of your plan.

   For instance, working with its dealers, Ford Motor began in-house development of its Smart Inventory Management System, or SIMS, which sought to equip dealers with data they could use to better predict which vehicles people would want to buy well before they set foot on the lot, allowing assembly plants enough time to make and ship them. That initiative worked toward satisfying more than one strategic objective for Ford.

4. *Take action.* At some point, thinking strategically must convert to activities. The late management scholar, Peter Drucker, framed the idea succinctly as: "All plans must deteriorate into action." Thus, a sense of proper balance is needed to look outward; yet, stay alert to the urgent marketplace and competitive needs to sustain the viability of the organization now and into the future.

5. *Align thinking with the corporate culture.* Any strategic thinking that reaches out in space and time must blend with the operating culture of the organization. That means: attempting forward-looking expansion to utilize disruptive technologies that aim to serve new customer groups, yet finds itself operating within a self-satisfying, complacent culture, is unlikely to reach its goals. Either the culture changes or the expansion objectives shrink. The impact of culture has been cited in various chapters.

6. *Communicate to all levels within your authority.* Much has been said about internal communications in Chapter 7. Holding a single

strategic viewpoint without exciting those with whom you work will not likely achieve reality. Continuing to strive for excellence by effectively communicating, motivating, and inspiring are qualities of excellent leadership.

Maneuver Tactically

A subset of thinking strategically, and its extension into turnaround planning, links to tactical maneuvers. These are defined as short-term actions to achieve long- and short-term objectives.

Such a move that blends the strategic and tactical is illustrated by Google. When the technology giant made one of its major updates to Android (the operating system that runs on 85 percent of the world's smartphones), it was done with little fanfare compared to the elaborate pomp that introduced competing iPhone 6 and 6 Plus.

At the time, Google introduced three newly designed devices, known as Lollipop, including the supersize Nexus 6 smartphone. It was part of a series of new Lollipop-powered computers for use in living rooms, cars, and just about everywhere else.

As a Google executive expressed the maneuver: "We aren't only trying to ship two products (referring to Apple's two models), we are trying to enable thousands." Clearly, Google was thinking strategically as it looked to control the next generation of computing devices, and it was maneuvering tactically in the day-to-day contested battle of matching wits and products with aggressive competitors.

How, then, do you prepare for tactical maneuvers? They begin by examining the fundamentals.

First, look at the makeup of your product mix and determine if there is sufficient depth and width. Apple decided on two iPhone models; Google introduced three devices, which was part of an ongoing series. Second, acknowledging the reality of the product life cycle,* review the scope of your new product development plans. Are product plans made up of tactical and superficial changes, or are they expansive and strategic, as in Google's plans to "enable thousands?"

One way to see the relationship of the two approaches is to view the categories of products. *New* means new to the marketplace and can take

---

* The product life cycle refers to the generally accepted stages of a product's life: introduction, growth, maturity, decline, and phase-out.

TABLE 8.1

Categories of New Products

Category	Definition	Nature	Benefit
Modification	Altering a product feature	Same number of product lines and products	Combining the new with the familiar
Line extension	Adding more variety	Same number of product lines, higher number of products	Segmenting the market by offering more choice
Diversification	Entering a new business	New product line, higher number of products	Spreading risk and capitalizing on opportunities
Remerchandising	Marketing change to create a new impression	Same product, same markets	Generating excitement and stimulating sales
Market extension	Entering a new market	Same products, new market	Broadening the base

many different forms. This diversity can be viewed by varying degrees of technological and market newness. In terms of increasing degrees of technological change, you may want to distinguish among modification, line extension, and diversification.

For increasing degrees of market newness, you can differentiate between remerchandising and market extension. Table 8.1 presents the differences between these five categories of new products and points out the benefits of each.

Pricing is still another way to maneuver. Some initial questions you have to consider are: What perception or image do customers hold in their minds about your product as it relates to low price or high price? Once you establish a price position, would you create confusion in their minds if you decide to change the fundamental image of your firm? Such a negative situation occurred when J. C. Penney attempted a total change in direction during its turnaround, as cited in Chapter 2.

Consequently, give careful consideration to these questions when introducing a new product and devising a pricing strategy. For instance, some organizations recognize image as a precious factor and will create a new name brand within a low-price category just to avoid conflict rather than run the risk of damaging the image of its upscale product.

In general, it is difficult to regain a premium price position for the same brand once it has been diluted by low-price promotions through mass merchandising outlets. Therefore, as you shape a strategy for a new product entry, it is wise to maintain ongoing feedback about the market position you want. In turn, the market position you select ultimately has consequences for your product's image.

The following represent types of pricing strategies:

- *Skim Pricing.* This approach involves pricing at a high level to hit the "cream" of the buyers who are less sensitive to price. The conditions for using this strategy include:
  - Senior management requires that you recover R&D, equipment, technology and other startup costs rapidly.
  - The product or service is unique. It is new (or improved) and in the introductory stage of the product life cycle, or it serves a relatively small segment where price is not a major consideration.
  - There is little danger of short-term competitive entry because of patent protection, high R&D entry costs, high promotion costs, or limitations on availability of raw materials, or because major distribution channels are filled.
  - There is a need to indirectly control demand until production is geared up.
- *Penetration Pricing.* This is a maneuver to price below the prevailing level in order to gain market entry or to increase market share. The conditions for considering this strategy include:
  - There is an opportunity to quickly establish a foothold in a specific market.
  - Existing competitors are not expected to react to your prices.
  - The product or service is a "me too" entry and you have achieved a low-cost producer capability.
- *Psychological Pricing.* This pricing strategy means pricing at a level that is perceived to be lower than it actually is, e.g., $99, $19.99, and $1.98. Psychological pricing is a viable maneuver and you should experiment with it to determine its precise application for your product. The conditions for considering this strategy include:
  - A product is singled out for special promotion.
  - A product is likely to be advertised, displayed, or quoted in writing.
  - The selling price desired is close to a rounded-out number of 10, 100, 1,000, and so on.

- *Follow Pricing.* This approach means pricing in relation to industry price leaders. The conditions for considering this strategy include:
  - Your organization may be a small- or medium-size company in an industry dominated by one or two price leaders and there is no opportunity to outmaneuver the market leader.
  - Aggressive pricing fluctuations may result in damaging price wars.
  - Most products in a given category don't have distinguishing features.
- *Cost-Plus Pricing.* This form of pricing means basing price on product costs and then adding on components, such as administrative costs, R&D expenditures, and profit. The conditions for using this strategy include:
  - The pricing procedure conforms to government, military, or construction regulations.
  - There are unpredictable total costs owing to ongoing new product development and testing phases.
  - A project or product moves through a series of start-and-stop sequences.
- *Slide-Down Pricing.* The purpose of slide-down pricing is to maneuver prices down to tap successive layers of demand. The conditions for considering this strategy include:
  - The product would appeal to progressively larger groups of users at lower prices in a price-elastic market.
  - The organization has adopted a low-cost producer strategy by adhering to learning curve concepts and other economies of scale in distribution, promotion, and sales.
  - There is a need to indirectly discourage competitive entries.
- *Segment Pricing.* This pricing maneuver involves pricing essentially the same products differently to various groups. The conditions for considering this strategy include:
  - The product is appropriate for several market segments, especially those that are poorly served, unserved, or emerging, or it is priced differently for such groups as students, adults, seniors, or military.
  - If necessary, the product can be modified or packaged at minimal costs to fit the varying needs of customer groups.
  - The consuming segments are noncompetitive and do not violate legal constraints.

- *Flexible Pricing.* This approach provides an opportunity to meet competitive or marketplace conditions. The conditions for considering this strategy include:
  - There is a competitive challenge from imports.
  - Pricing variations are needed to create tactical surprise and break predictable patterns.
  - There is a need for fast reaction to outmaneuver competitors' attacking your market with penetration pricing.
- *Preemptive Pricing.* This pricing strategy indirectly discourages competitive market entry by being first to introduce a product at the same or lower price than a rival's expected entry into the market. The conditions for considering this strategy include:
  - You hold a strong position in a medium to small market.
  - You have sufficient coverage of the market and sustained customer loyalty to cause competitors to view the market as unattractive.
- *Phase-Out Pricing.* This kind of pricing means pricing high to remove a product from the line. The conditions for considering this strategy include:
  - The product has entered the downside of the product life cycle, but it is still used by a few select customers.
  - Sudden removal of the product from the line would create severe problems for your customers and create poor relations.
- *Loss-Leader Pricing.* Pricing a product low to attract buyers for other products is known as loss-leader pricing. The conditions for considering this strategy include:
  - Complimentary products are available that can be sold in combination with the loss leader at normal price levels.
  - The product is used to draw attention to a total product line and increase the customer following. The strategy is particularly useful in conjunction with impulse buying.

Your overriding purpose in all of the above strategies is to avoid price wars. Rather, the intent is to use pricing to indirectly enter untapped market segments and focus on product improvements. You also can preempt and discourage new competitors by gradually sliding down prices, thereby making the market seem unprofitable. Also, you can always maneuver with price according to the flexibility of demand and your production economies.

Marketing is a third approach to tactical maneuver. It incorporates all forms of communications, sales promotion, and sales, accentuated by the huge shift to the Internet and social media. Specifically, the central issues include adjusting the communications mix to the optimal use for stimulating sales and neutralizing competitive actions; organizing the sales force to provide market coverage to meet customers' logistical, technical, and service needs, and for enhancing the effectiveness of the supply chain.

The following review deals only with the primary marketing considerations related to advertising and sales promotion. (Extensive coverage of employing the sales force is beyond the scope of this section.)

Tactical Advertising. Effective advertising is meant to result in positive attitudes and behavior. Thus, there is the subtle application of indirect strategy, which is to influence the receivers of the message in such a way that it results in increased sales. Yet, to say the objective of advertising is to increase sales is far too broad an outcome for an advertising program.

Rather, you, or those responsible for implementing advertising campaigns, should formulate more specific and measurable aims that precisely pinpoint what is expected from advertising. For example:

- Support the sales force.
- Achieve a specific number of exposures to your target audience.
- Address prospects that are inaccessible to your salespeople.
- Create a specified level of awareness, measurable through recall or recognition tests.
- Improve relationships along the supply chain.
- Enhance consumer attitudes toward your product or company.
- Launch a new product and generate demand for it.
- Build familiarity and easy recognition of your company, brand, package or trademark.
- Counter false or inaccurate claims made by competitors.
- Solve a problem.

The list illustrates some of the possibilities and identifies the need for precision when setting advertising objectives. Because objectives imply accountability for results, they often lead to a performance evaluation of an individual, marketing department, or advertising agency.

Table 8.2 details the steps involved in developing an advertising campaign to support an indirect strategy. It shows that continuous market and competitive research is the foundation of a sound campaign and plan.

TABLE 8.2

Developing an Advertising Campaign to Support an Indirect Strategy

Precampaign Phase

1. Market and competitive analysis		Study competitive products, positioning, media, distribution, and usage patterns
2. Product research		Identify perceived product features and benefits that represent areas of differentiation
3. Customer research		Conduct demographic and lifestyle studies of prospective customers, investigate media, and purchasing and consumption patterns

Strategic Decisions

4. Set objectives	Determine target markets, evaluate competitors' strengths and weaknesses, and identify user profiles	Develop metrics of evaluating performance
5. Decide on budget	Determine total advertising spending necessary to support objectives	Investigate competitive spending levels and media costs necessary to reach objectives
6. Formulate advertising strategy	Develop creative approach and prepare media list	Examine audience profiles, reach, frequency, and costs of alternative media
7. Integrate advertising strategy with marketing strategy and the overall turnaround plan	Make sure that advertising supports and is supported by other elements of the marketing mix	

Tactical Execution

8. Choose advertising themes, message, and mode of presentation	Develop alternative creative concepts, e.g., employing social media and use of evolving technologies	Conduct concept, theme, and message tests
9. Establish final media plan	Determine media mix and schedule	Conduct media research, primarily from secondary sources
10. Campaign implementation	Finalize campaign details with agency or marketing department	

Campaign Follow Through

11. Impact control	Obtain feedback on consumer and competitive reaction	Evaluate performance based on advertising objectives and other metrics of performance
12. Review and revision	Adjust media, content, frequency, and spending based on campaign objectives and market conditions	

Sales promotion is another tool of marketing. The idea is to integrate sales promotion with your advertising and sales force objectives and strategies. Whereas advertising offers a reason to buy, sales promotion is an incentive to buy. Also, while sales promotion is part of an overall tactical program, it involves a variety of company functions to make it work effectively.

Sales promotion permits tremendous flexibility, creativity, and application. Consequently, sales promotion is not a stand-alone activity. Instead, it is a vital component of tactical maneuvers. As with advertising, it is best employed when linked to the broader vision of your plan's strategic direction and when tied to specific objectives, such as:

- Entering new market segments
- Gaining entry into a new supply chain
- Encouraging upselling to increase revenues per customer
- Building trial usage among nonusers
- Attracting prospective customers from competitors
- Building brand loyalty
- Stimulating off-season sales
- Retaining existing customers and winning back former customers.
- Outmaneuvering competitors through preemptive promotions and neutralizing their actions

Sales promotion includes an array of activities from demonstrations and contests to coupons, premiums, and samples. They are directed at one or any combination of three distinct audiences: a company's own sales force; middlemen of all types and levels, such as wholesalers and retailers; and end-use consumers or business-to-business buyers. Specific applications include the following:

1. *Motivating the Dealer.* With dealers (or any intermediary in the business-to-business, consumer, and service sectors), the most powerful language to speak is still money, i.e., profit. Among many available sales promotion techniques to motivate dealers are buying allowances, cooperative advertising arrangements, dealer listings, sales contests, sponsorship of special events, and support at trade shows.
2. *Introducing New Products.* Sales promotion techniques particularly well suited to the introduction of new products include free samples or trial offers, coupons, money refunds, displays, workshops, and

presentations, to name a few of the most commonly used approaches. And, then, there is the pervasive use of social media to create impact and sustain the effort.
3. *Promoting Existing Products.* Several tools are available to promote, and re-promote, established brands, such as premiums, discounts, contests and sweepstakes, and demonstrations. These methods aim to attract competitors' customers and build market share, introduce new versions or product-line extensions of established brands, and reward buyer loyalty.

The central focus is to employ these tactical tools to activate indirect strategies and provide flexibility for maneuver. Further, the aim is to avoid resource-wasting, head-on clashes with competing companies.

Unbalance the Competitor

The third underlying principle of indirect strategy is combining the physical with the psychological to create an unbalancing effect on the rival manager. The physical force was described above. The aim of the psychological force is to create diversions or distractions through maneuvers, devices, feints, and even misinformation.

Doing so creates a dislocating effect within the opposing executive's mind that influences the individual to look in other directions and arrive at inaccurate decisions about the actual situation. The psychological effect is created in a variety of ways. For instance, Bloomingdale's, the upscale department store, uses technology to help solve a common problem that has cost apparel retailers deeply in sales; that is, irritated customers leave the store rather than take the time to get hold of the right piece of clothing.

Bloomingdale's solution was to install "smart" fitting rooms equipped with wall-mounted iPads. The customer or staff member can rapidly scan the item in question to find which colors and sizes are in stock. The tablets then permit customers to see ratings and reviews by other customers. It also recommends items that would complement the scanned original.

There are additional benefits to the system that can be termed *indirect strategy*. The technology identifies and uses unsold inventory from physical stores to fill online orders rather than sell it on clearance. It also offers store pickup for orders placed online. The strategic idea is that the technology helps reduce the loss of customers to other e-commerce competitors.

What are the central issues? First, Bloomingdale's had a perennial problem that it needed to solve, which was to relieve customer frustration during the trying-on and purchasing process. Further, the retailer needed to overcome intense competition from other physical stores and online retailers. Its indirect strategy was to outmaneuver competitors by integrating technology into the shopping experience.

The question, then, is what effect would those innovations have on the minds of competing managers? While personalities vary, there is a psychological impact of some sort, typically exhibited by behavioral factors, such as stress and fear that contribute to an unbalancing effect among individuals confronted by a sudden or unexpected development.

Stress

Stress is a reaction to events that causes an individual to feel endangered or vulnerable, depending on the nature of the event. Whether real or imagined, the body responds with a reaction that affects the mind, body, and behavior in many ways.

From a psychological viewpoint, the symptoms can appear as poor memory, inability to concentrate, poor judgment, excessive worrying, moodiness, agitation, or the feeling of being overwhelmed. In other words, these symptoms in varying degrees can create an unbalancing effect on a manager and, consequently, on his or her decision-making capability. In turn, it has an escalating effect on total performance.

These visible shifts in behavior also tend to trickle down from the affected leader to members of the staff who can easily see the changes in mannerisms. As an outcome, the leadership qualities of objectivity, calmness, optimism, strength, and clear thinking are marginalized.

On the other hand, there is the possible condition where a "fight-or-flee" reaction surfaces and the manager tends to fight rather than flee. Such is the case of an Arizona utility operating in the rare business world of a market monopoly and enjoying a regulated 10-percent return on its invested equity. Then, the executives in that organization felt threatened by the raging growth of solar energy through firms such as SolarCity. Noticing the lost revenues from customers who switched to solar energy, the utility sought government regulations to impose taxes on solar installations.

The essential point being that indirect strategy remains a potent psychological force. To illustrate the far reaches of this concept from another field of conflict, Napoleon quantified the effect as: "The psychological is to the physical as 3 to 1."

As with the application of any strategy, there is the human factor whereby emotions make up a major part of an individual's thinking. Expressed in a more pragmatic way: In a competitive encounter, the mind of one manager is pitted against the mind of a rival manager. Therefore, it is in your best interest to maintain a vigilant, top-of-mind awareness of your competitor.

Fear

Fear takes several forms: fear of death, fear of survival, fear of the unknown, fear of uncertainty and unpredictability. It is the latter three that are most common to business managers. As with the symptoms of stress, it depends on the type of personality and how an individual reacts to a particular incident.

Uber, discussed in the beginning of this chapter, in attempting a competitive advantage by using its car-financing program as an indirect strategy, created a psychological advantage over its competitors, such as the owners and managers at Lyft and local taxi companies, as well as their drivers, many of whom ending up moving to Uber.

Fear, then, is a deeply rooted part of human nature. Such an emotion would certainly have entered the minds of those rivals in varying degrees as Uber's strategy took its effect. Again, the power of unbalancing the competing manager should be a major consideration in developing any type of indirect strategy.

Thus, indirect strategy consists of an interlocking of physical and psychological forces. Further, the object is to circumvent direct confrontation, because competitive fighting is not the object of strategy. It is achieving a business objective, such as occupying a profitable, long-term position in a market.

Nor is sustaining excessive losses of capital and other resources by fighting the mark of a successful executive. It is creating relationships and consummating a sale among customers as it fends off threats from competitors—indirectly.

Referring again to other fields of endeavor where rivals are contesting ground or fighting for a cause, there are clear-cut references to indirection that hold a large measure of wisdom:

184 • *Developing a Turnaround Business Plan*

For to win one hundred victories in one hundred battles is not the acme of skill. To subdue the enemy (rival) without fighting is the acme of skill. Supreme excellence consists in breaking the enemy's resistance without fighting.

Sun Tzu

If we always knew the enemy's intentions beforehand, we should always, even with inferior forces, be superior to him.

Frederick the Great

In sum, an indirect strategy applies strength against a competitor's weakness, resolves customer problems with offerings that outperform those of your competitors. As for your competitor, your aims are twofold.

First, achieve a psychological advantage by creating an unbalancing effect in the mind of your rival manager by distracting him or her into making false moves and costly mistakes. Second, create approaches as described in this chapter that neutralize the rival's capabilities that would prevent you from achieving your objectives.

Next, integrate indirect strategies into your turnaround plan by engaging colleagues and staff, ideally through the diversity of a cross-functional team, to open their minds to fresh ideas about innovative ways of developing such maneuvers. You thereby reduce the risks and increase the chances of success of going after market leaders, even where limited resources are available.

As you develop your plan, look for approaches that would probe for unserved market niches where there is minimal resistance from larger competitors, and where opportunities exist to establish a foothold and expand into a mainstream market.

Finally, much of what is discussed in this chapter is illustrated by a relatively small firm in a big industry that has gone through massive changes: the book business and, specifically, book stores. By far the likes of Amazon have had the greatest disruptive effect by upending the way people buy books.

Then, there are the disrupting changes in the way people read books— think e-readers and tablets. All have contributed to the closings of many small book stores and even the once mighty Borders.

Within that maelstrom of powerful forces, one book store company, Half Price Books, does exist and from all indications operates profitably, debt-free, and with continuing signs of growth. The company found a market niche in which to concentrate and uses an indirect strategy through which to function in a predominantly online buying marketplace.

The niche consists of those individuals who like to browse bookstores and take delight in what they find. And, there appears to be substantial numbers of those customers to support its 120 retail locations in 16 states, as of 2014. A survey conducted by Half Price Books revealed that its customers bought on average 37 books a year. Armed with that data and other intelligence, CEO Sharon Anderson Wright plans to continue the company's expansion by following its pattern of adding five new stores each year.

What, specifically, is behind Half Price Book's indirect strategy?

Wright keeps its real estate costs low. That approach is in contrast to the likes of Barnes & Noble that is locked into new books and high-priced locations. She buys virtually anything that is printed (or recorded), excepting newspapers, by sending five to six buyers around the country to purchase remainders that can sell at half price. The company also produces its own stationery, calendars, and CD wallets. And, through its wholesale division, sells to museums, independent bookstores, and even Barnes & Noble.

Consolidating all the above factors permits Half Price Books to remain nimble by not creating direct confrontation, carving out a viable niche, targeting its audience, and controlling fixed costs. It thereby prospers as it maneuvers in a dynamic marketplace through an indirect strategy.

9

Valuing Surprise and Speed

Chapter Objectives

Be able to

1. identify the advantages to employing speed in a campaign;
2. describe the two primary purposes for initiating surprise in a competitive campaign;
3. recognize the barriers that prevent implementing surprise;
4. develop a positioning strategy; and
5. determine the ending point of a campaign.

INTRODUCTION

The previous three chapters dealt with bold action, concentration, and indirect strategies. Surprise and speed now become the propellants to activate turnaround strategies. What is the underlying meaning and value of surprise?

Taking a competitor by surprise has two central purposes. First, to achieve superiority at a decisive point, which is one of the foundation aims of most campaigns; and, second, to neutralize the rival's capabilities that would prevent you from achieving your objectives. If looked at from those vantage points, success without surprise is more difficult to achieve without getting involved with more intense competitive skirmishes and excessive expenditures of resources.

Surprise, therefore, is the means to gain a pivotal advantage. One of its greatest effects is the psychological impact of unbalancing the opposing manager, a major issue discussed in the previous chapter. Whenever it is achieved on a grand scale, it extends into the rival organization and dampens the morale of a group or seeps into additional layers of personnel.

Consequently, surprise should be made part of all campaigns without exception, though in varying degrees depending on the nature and circumstances of the operation. Circumstances affecting how surprise would be implemented include the rival organization's structure, the layers of management used in decision making, the sophistication of the internal communications network, the morale of the individuals involved, the quality of the planning effort, and certainly the effectiveness of the leadership and its ability to develop counterstrategies.

The additional factors that produce surprise are secrecy, security, and speed. Both presuppose a high degree of energy and discipline on the part of the attacking organization and its leader. Surprise will never be achieved under lax conditions and conduct.

Although the dominant aim is to achieve total surprise, it is somewhat unrealistic to think that absolute surprise can be achieved. The reasons being that the variables listed above do exist, as well as the individual practices of each competitor, such as how vigilant is the staff, how effective is the intelligence-gathering system, and how alert are managers in reading the signs of danger. The following case illustrates one dimension of surprise and its remarkable outcome.

Samsung, the South Korean company, was caught off guard. The company held a longtime leadership in smartphones in India and China. Then, in 2014, it lost no. 1 positions in both markets in a matter of months. The primary cause was Xiaomi.

With astonishing speed, the Chinese company put its Redmi 1S smartphones up for sale in India, using local e-commerce site Flipkart. In just four seconds (that is the reported number), 100,000 phones sold out at $98 apiece. That pattern had become a weekly event within three months of entering India.

If employed effectively, then, surprise can be a key element of success in your turnaround plan. That is, as long as you can make it work. As indicated, surprise requires total security of your plan, which could mean issuing misinformation to deter competitors from knowing your real intentions.

Admittedly, to some extent that act can border on deception. However, it may be necessary, within legal bounds, to provide some seal of security.

As for the sharing of information about plans, dates, and strategies, these should be handled with the greatest care and discretion.*

Then, there is the current priority of dealing with the threat of cyber attacks. As of 2013, of more than 500 U.S. businesses, government agencies, and law enforcement services that responded to a survey, only 38 percent said they strategically invest in cybersecurity, and only 17 percent reported taking steps to identify which business data are most important.

Other findings from that survey reveal a lack of attention to the security practices of contractors, supply chain partners, and other third-party business partners. Less than half of the group surveyed reported having a process for evaluating third parties before they launch business operations. Less than a third included security provisions in contracts with external vendors and suppliers. Then, an additional troublesome issue emerged that revealed less than half of respondents admitted that they did not offer security training to new hires.

Still another significant matter surfaced that related to blocking the implementation of surprise: friction within the organization. As was pointed out in Chapters 3 and 5, friction stems from multiple sources, such as the resistance resulting from organizational logjams that prevent the clear communication of directions from senior management to field personnel. It also can be failure to gather and organize accurate intelligence in a useable format. Thus, friction occurs when decision-making managers are unable to correctly estimate the situation and act rapidly to find a correct pathway for surprise.

Then, there is the problem of inexperienced or inadequately trained staff. Those individuals are the ones with the responsibility for implementing offensive strategies, which require discipline, cooperation, commitment, and mental alertness to stay balanced against the ups and downs of competing against aggressive rivals.

Friction, then, is pervasive. It fosters errors and places barriers in front of efforts to create surprise. With the sources of friction seemingly unlimited, you should be fully aware of the ones that can cause the greatest harm in your firm, then do what is necessary to limit the damage.

In practice, surprise is a tactical device that is more accurately measured by the precise dimensions of time and space. Where it applies to strategy, surprise becomes more feasible the closer it occurs to tactics. That is, it is

---

* Competitive intelligence and the use of live agents were covered in Chapter 6.

more difficult to surprise a competing organization when attempting to affect change associated with long-term strategic issues.

The scale of time and space was made clear in the Samsung example, above. Also, surprise is more easily carried out in campaigns requiring a limited timeframe.

Another characteristic of surprise is that the greater the ease with which surprise is achieved, the smaller is its effectiveness, and vice versa. In the abstract, you might believe that small surprises often lead to greater things, such as a major victory, but the history of competitive campaigns does not bear this out.

The general rule is that the weaker the forces that are at your disposal, the more appealing the use of cunning becomes, and, if in a state of weakness where prudence, judgment, and ability no longer suffice, cunning may well appear the only hope.

Further, the more urgent the competitive situation, with everything concentrating on a single campaign, the more readily cunning is fused to boldness. If you are in a state where you are free to make bold moves, there is the likelihood that your chance of success will increase.

Therefore, major success in a surprise action does not depend entirely on the energy, forcefulness, and resolve of the leader. As was pointed out above, it must be attributed to other circumstances, such as how embedded the competitor is in its position and how likely the rival manager can be dislodged from that position.

One more observation needs to be made, which goes to the very heart of the matter. If you surprise the competitor with faulty approaches or insignificant areas of differentiation, you may not benefit at all, but, instead, you may suffer sharp reverses. Your surprise, in that case, will have little or no meaningful effect and cause the rival little worry. Then, too, there is always the possibility that he would exploit your mistake to his own advantage.

The offensive, then, offers many more options for achieving surprise than does the defensive. Consequently, the element of surprise is more often related to offensive campaigns.

Should you benefit from the psychological effects of surprise, the better the outcome, especially when the rival manager finds he is incapable of making coherent decisions. In that instance, you can intimidate and outdistance the other by using surprise to greater effect and may even reap the fruits of victory where ordinarily it might fail.

What, then, can interfere with achieving surprise? There are several variations in the characteristics of the organization and certainly in the leadership.

A prime one is deficiency in planning skills. Specifically, that includes the failure to utilize a cross-functional team of cooperative individuals who can develop an effective, competitive strategy plan.

However, a fundamental problem that often impedes constructive output from the team is that some members may be inadequately trained in such strategy principles as speed, boldness, concentration at the decisive point, indirect strategy, maneuver, and the other elements that permeate the chapters of this book.

Further, surprise takes skill, knowledge of the marketplace, and, above all, awareness of the competitor's behavior. For instance, what will be the competitor's overall reaction during periods of sudden market activity and immediately after the surprise? Will it be reserved or aggressive? Offensive or defensive?

Then, from your vantage point, is there an effective communications system from the field to the home office? Does such a system exist for developing a postsurprise follow-up?

Finally, there is the all-encompassing issue of morale. Surprise needs a committed and spirited group of individuals who can react quickly and decisively to exploit the unfolding events that follow a campaign anchored to surprise.

SPEED

"Without exception, all of my biggest mistakes occurred because I moved too slowly," declared John Chambers, CEO of Cisco Systems. A voice from another time and field of human endeavor, Napoleon stated, "The whole art of war (conflict) consists in a well-reasoned and extremely circumspect (indirect) defense, followed by a rapid and audacious (bold) attack."

Those statements translate to several truisms about the use of speed in your operations.

First, drawn-out, slow-moving campaigns have rarely been successful. The central reasons is that those employees directly involved with the effort are usually in a high state of energy and expectations when the campaign begins. Where favorable results are not initially attainable, they are not able to sustain the physical energy and aggressive state of mind necessary to continue the campaign with the same zeal. Consequently, their half-hearted interests are diverted, morale is depressed, and budgets depleted.

Then, there is the problem of a cumbersome organizational framework where extended deliberations at the executive level and an inability to make timely decisions also slow events to a crawl until time wears away at the opportunity. Consequently, the lean organization is the trend for greater efficiency and speed, and flattening the organizational layers result in improved speed and increased flexibility.

In turn, a more flexible organization can achieve greater market penetration because it has the capacity to adjust to varying circumstances more rapidly. Therefore, it can concentrate at the decisive point before competitors have a chance to respond.

Speed, then, is acknowledged as an essential component to secure a competitive lead, with a positive impact on product position and customer relationships. Where a bold approach is used against a competitor's inroads, speed of response is an essential means to achieve a positive end.

As for customer relationships, the key driver is customer engagement. It is all about you and your staff being aware of the nuances and evolving patterns of consumer behavior. It is also for you to unearth insights from data, and now Big Data, and apply them to a diverse range of interactive activities and exchanges with customers.

Where there is a flood of minor or disruptive technologies being introduced rapidly, you and your staff must know that a vibrant strategy that integrates speed with technology is in a prime position to secure a competitive lead. Thus, speed adds vitality to the operations and acts as a catalyst for growth.

Second, any delay gives watchful competitors time to mobilize enough effort to put up a strong defense to oppose you, or it can be a reason to go on the offensive against you. "None of this is all that proprietary or genius. That's why I feel like I have to run like the wind," announced Michael Dorf, the head of City Winery, a chain of music venues that offers food, wine, and rock 'n' roll in an intimate setting. He senses how fragile his lead can be and how easy a competitor with a sound plan can outperform him during any lax moment.

Should there be a loss in momentum, any attempt to recover market share, competitive position, and customer loyalty are often more costly, time consuming, and riskier than moving swiftly at the first signs of slowdown.

Market share, in particular, is a fundamental measure of how well you are doing in the marketplace compared to competitors. There are numerous factors to measure, so that it is meaningful to use similar metrics when benchmarking against another firm.

When looking at market share, take into account the following factors:

- Total market size as measured by annual growth and rate of consumption, usually tracked over a period of three to five years.
- Market share, which is the size of the market expressed as a percentage against (1) your entire industry, (2) the top competitors that directly impact your operations, and (3) the single most important competitor you currently face.
- Market definition, which is determined by such factors as price, quality, speed of service, ease of maintenance, points of distribution, and any factors pertinent to your business. The intent is to provide more precise targeting information and to identify segments that are emerging, neglected, or poorly served. The definition also includes an analysis using such criteria as demographics, psychographics, and product usage data, which add greater precision in understanding the nuances that make up market share.

Third, speed impacts a product's sales life as product life cycles continue to contract at ever-increasing rates. In many instances, it is the result of the rapid introduction of new technologies and their early adoption by swift-moving competitors.

Such is the case of Square, an innovative high-flyer with its little card reader attachments for phones and tablets. In only four years, the Silicon Valley start-up built an infrastructure of card readers from scratch, especially through small business owners.

As of 2014, however, the rapid rate of growth came to a screeching halt. More than 80 companies, including Apple, PayPal, and Amazon.com, offered card readers that looked a lot like Square's, often for less and with more features. And, technology continued to move forward with rapid changes as merchants switched to checkout terminals that can accept cards using chips instead of magnetic stripes.

By anticipating the eventual commoditizing of an existing product—and subject to the varying influences that affect speed—product development is a factor of timing, which should begin at the growth stage of its product life cycle. From a strategy viewpoint, the important considerations in the product development process are positioning and branding.

POSITIONING

Where positioning has a physical dimension as to where and how a product is placed in the market, there is also the all-important position the product holds in the customer's mind. That is, the product has to be placed with a sense of the emotional satisfactions the product offers and the image it casts. What follows is to monitor those forms of fulfillments and perceptions through observation and research. Doing so, defines the position in the market and in the customer's mind.

Pioneered by Al Reis and Jack Trout during the 1980s, positioning was popularized as: "Not what you do to a product. Positioning is what you do to the mind of the prospect."

And from Professor Philip Kotler (Northwestern University): "Positioning is the act of designing the company offer and image so that it occupies a distinct and valued place in the target customers' minds."

Both definitions presuppose that a product's position results from securing a valued place early on in the mind of the prospect and, barring unforeseen circumstances, retaining that dominant position.

Related to the overall concepts of positioning, consider the following guidelines:

Maneuver. Position your products in those niches where there is an above average chance to gain a rank placement among the leaders. Where possible, avoid the commodity segments, unless you are a price leader and that represents the driving force of your entry strategy. Ideally, find a technology, product design, supply chain, or service that differentiates your product or company and leads to a favorable position compared to that of competitors.

Establish flexible work teams. Cross-functional teams create the vital linkage between customer and successful product development and positioning. To succeed, however, teams must have a clear definition of how the company wants to be positioned and be able to implement the desired position within the framework of the turnaround plan.

Solve customers' problems. The extent to which you are able to solve customers' problems also contributes to establishing a solid position. Therefore, look for new product applications, value-added services, and new market segments that were overlooked in the initial stages of product development to solidify your position.

For instance, one problem that plagued retailers was watching frustrated shoppers walk away without purchasing rather than endure the long check-out lines during high traffic hours. The owner of one retail chain, In the Pink, addressed the problem by turning to technology and making his store personnel into mobile cashiers. The traditional checkout counters were ripped out and replaced with racks of merchandise. Store clerks, armed with iPad minis, roamed the stores and intercepted customers either in the process of making a selection or when exiting the dressing room. The clerks quickly pulled up a customer's history, checked inventory in the other stores, and completed the sale. Four months after moving to mobile point of sale, the retailer witnessed a 23-percent jump in same-store sales compared with the same period a year prior.

Look globally. Trade barriers and various forms of government restrictions vary with location. Where you have some freedom of movement, gain a foothold and push your product ideas and technologies wherever they apply in the world through acquisition of local companies in the particular area, or by various forms of joint ventures.

In local markets follow the principles indicated above. That is, make sure you are positioned to satisfy local needs with customized products and services, and not use foreign markets as a means to unload a standardized product.

Develop a Positioning Strategy

If the picture of your market reveals an undesirable position for your company or brand, use the following procedure to help improve your situation:

- Identify your product's actual position, which requires individual consumer interviews, generally in the form of surveys.
- If you accept your brand's current position, make certain it commands a strong position in its field and there is persuasive evidence of continuing growth potential. Another approach is to select a position that nobody else wants, that is poorly served, overlooked, or emerging.

- In attempting to achieve an ideal product position, your firm has two principal options. It can (1) move its current product to a new position, with or without a change in the product itself; or (2) introduce another product with the necessary characteristics for new positioning, while leaving the current product untouched, or possibly withdrawing it from the market.
- Once you discover that your product's position is far from ideal, your marketing effort takes on a formidable job of making the shift to the new position. Together with the other elements of your communications mix—namely, personal selling, publicity, social media, the Internet, and sales promotion—your advertising will have to shoulder the burden of creating a new position for your product.
- After developing several alternative strategies for achieving your ideal product position, select one of them to implement in the marketplace. In making your decision, be guided by your company's overall objectives, resources, and capabilities. Consider, too, how long and how firm a commitment your company is willing to make, and how much money it is ready to put behind such a commitment.
- Achieving a lasting and favorable position is an expensive, time-consuming proposition. Unless your company's management is firmly committed to this strategy, it is best not to tamper with your brand's position, unless it is absolutely essential to survival. You might do more harm than good if the effort is half-hearted or is terminated halfway into the program.
- While tracking your competition, monitor the impact of your positioning on the customer's mind where it counts most. Follow-up research must examine and compare your product's actual position with its desired ideal position. After all, it is possible that your program will not produce the intended results. In this event, a review of your strategy may be necessary.

The primary goal of positioning, then, consists of a two-pronged strategy: (1) create a long-term desirable position for your product in the customer's mind and (2) secure a strong advantageous position against your competition.

Branding

Allied with positioning is branding, which conveys to your customer what they can expect from your products or services, as well as how it is distinguished from your competitors. Your brand is a reflection of your company's core values and culture. It is communicated by means of your logo, website, packaging, promotional materials, and the people representing your brand. All of these modes deliver your brand image.

The intrinsic value of your brand, thus, is expressed in the form of perceived quality and its emotional appeal. BMW and Chevrolet each has a distinct branding image that targets a defined segment of the population with its related associations of people and events.

Therefore, defining your brand encompasses a variety of factors and requires a series of steps.

First, develop a distinctive logo and display it extensively among those groups or segments that you targeted as a primary audience. Tell a meaningful story that conveys the feelings or emotional appeals you want to express. After you determine if your message resonates with your prospects, keep repeating the basic themes of your story. For example, the design of a new line of watches and driving shoes captured the story of the Mini Cooper's cool, urban driving experience. The happening related to the driver, not the car.

Second, integrate your brand into the trappings that are visible to your intended audience. The shabby display at a key trade show staffed by inappropriately dressed representatives hardly offers a quality image.

Third, consistency, then, is one of the major factors in the branding process. One central aim is to avoid confusion in your customers' and prospects' minds. Therefore, make sure your brand has a timeframe that conforms to the long-term strategic direction of your business plan.

Fourth, speed adds spirit to a company's operations and serves as a stimulus for growth—and turnaround. The impetus elevates employee morale and tends to energize the group to move forward. Such energy builds into a meaningful force through the sheer momentum of time.

As an example, Southwest Airlines drove forward with a three-pronged strategy built around low fares, one free checked bag, and no change fees. As a result, as of 2014, it had become the nation's largest carrier of domestic passengers, beating out formidable rivals like American Airlines, United, and Delta.

As other airlines struggled through bankruptcy filings in recent years, Southwest kept its low fares and finances intact. The airline boasts four straight years of revenue gains. From a once regional airline, it started to expand internationally. With that expansion, management was concerned that a confused message would be sent to loyal customers that the organization was getting too big and unwelcome changes were about to take place.

Southwest decided its basic story had to be told and retold. To that end, the airline initiated a new corporate rebranding campaign that consisted of repainting its fleet of aircraft.

It went on to revamp everything from its cocktail napkins to employee uniforms. In addition, the big innovation was to paint the belly of each newly painted plane with a large heart. The idea was to reinforce the theme of the airline's basic story: "New look, same heart."

Even where an organization is defending its position, swift action is an active component, that is, if you accept the notion that defense is but one phase of a two-part process that allows for a pause only to regroup and then to move against a rival.

BARRIERS TO IMPLEMENTING SPEED

Even where the benefits of swift action are acknowledged, there are barriers that continue to slow well-intended plans to a disturbing crawl. It is in that work environment that tempers flare, frustration heightens, and general negative behaviors surface. It is in these surroundings that you have to locate the areas of friction and, to the extent possible, clear out the problem or override it.

The following guidelines will serve to pinpoint the most blatant obstacles:

Leadership

Where executives and managers can solve problems, they also can be the cause of problems that prevent implementing speed. The range of possibilities are vast and include such factors as a leader's basic personality that requires him or her to check and quantify every detail as time ticks away and there is only a limited period in which to take advantage of an opportunity.

Feeding that behavior may be inadequate or faulty market intelligence in which the leader has little confidence about its accuracy. Then, there is the case of sheer procrastination, whereby the overly cautious leader simply won't move, even though he or she knows that the needed data will not likely be available.

Beyond personality, there is another dynamic whereby a leader has risen in the organization with a dedicated specialty in which he has excelled and, thereby, had been promoted. However, when thrust into a situation that relies on the broader knowledge of markets, strategy, and maneuvers, the experience and training that feed the decision-making process is lacking.

There is a further twist to the above behavioral condition. The leader, for personal and career reasons, may feel reluctant to make a decision for fear of making an error. In turn, there is a feeling of insecurity about what is ahead, and even should these persons realize the urgency of speed, the more they linger with the potential dangers, the more doubtful and indecisive they become.

That all leads to a dominant leadership trait that was discussed in Chapter 4: *courage*. Although a reasonable degree of intellect is vital for a leader to grasp and internalize key factors of a situation, and there are certainly an ample number of brilliant executives in organizations, some may not have that other vital ingredient of being able to act with speed and determination. It takes an individual who can arouse the inner feeling of courage and the motivation to move at the critical moment when reason is pushed aside and replaced with action.

The Organization

Several conditions can create barriers to implementing speed. First, a general malaise exists that permeates the various groups and results in missed opportunities. Personnel seem to lack initiative in implementing business plans with any semblance of urgency.

Second, the problem may be caused by a sluggish and ponderous corporate culture that tends to hesitate and vacillate. From a corporate point of view, this is difficult to change unless it is attempted at the C-suite level.

At the group or product-line level, however, where there is some autonomy, change is possible by an assertive manager who can rally the group through a series of steps, such as seeking their full cooperation, developing plans and strategies through collaboration, and encouraging open communications.

Also, from an organizational structure viewpoint, breaking down the barriers by eliminating long chains of command aids in speeding up decisions. As an example, Royal Dutch Shell endured a sluggish culture at one time due to its joint British and Dutch management structure, which plodded along with two chairmen and two executive committees. It finally took the courage of a new CEO to streamline the organization. One immediate priority was to speed up the overly analytical culture, which made it difficult for the company to land big deals in a timely fashion.

The Ending Point

Speed tends to carry along with it a particular momentum. Once in motion, the tendency is to keep the campaign moving and then continue to move seemingly without looking back. Yet, there are two considerations to take into account.

First, there is a point where you have to go from the offensive to the defensive. The aims are to protect what you have achieved, consolidate your gains, and firm up customer relationships all along the supply chain. There is also the need to prepare against the inevitable efforts by competitors attempting to retrieve customers or recover lost market share.

Second, you have to know when to cease the offensive effort; that means determining the ending point of a campaign. Success in a campaign most often results from the concentration of superior strength at a decisive point, and, if the concentrated strength leads to a successful outcome, such as the possession of a market segment, the objective will have been attained.

However, the reality is that over an extended period of time strength diminishes as resources and budgets are depleted due to the continuous effort. The object, then, is to avoid getting your firm or group into a state of exhaustion.

There are some campaigns that have led directly to the stated objective of the plan in an on-time, lock-step approach. However, these are in the minority. Most campaigns only lead to the point where the available resources and budgets are just enough to stay marginally involved.

Beyond that point, the scale turns and the expenditures in time and effort don't produce a significant effect. In fact, continued attempts often become counterproductive. This is the essential meaning of the ending point of the campaign.

Thus, the ending point should be of vital interest to you, most senior executives, and line managers. This is especially so among those with decision-making responsibilities for shaping strategies and committing their organization's resources in campaigns that involve entering new markets against entrenched competitors.

Therefore, your task is to determine what would represent the ending point in a campaign. Would gaining an extra point of market share be worth the expenditures? Or would continuing the efforts and expending more resources place the entire campaign in doubt and result in exhausting financial, material, and personnel resources?

There are numerous factors that contribute to a decision to halt offensive activities. What matters, therefore, is to detect the correct ending point with discriminative judgment.

The ending point can be determined by using quantitative calculations related to the plan's objectives, and by compiling a list of nonquantitative criteria, such as assessing the morale of staff and the likely reactions of competitors to continuing the offensive, or it could be guided by appraising and prioritizing various opportunities highlighted in the turnaround plan. In the end, however, an answer may come to you intuitively through an inner voice that says, "It's time to end the campaign."

You also can decide the end point by calculating the sources of strengths and losses.

The causes of gains in strength include:

1. When the losses to the defending organization are greater than those of your firm (assuming you are the attacker), then it is to your gain.
2. The defender loses market share, and possibly its dominant position, from the time you entered the rival's market space.
3. You benefit from strengthening your position in the supply chain.
4. The rival loses staff cohesion and the smooth functioning of its marketing force.
5. Some external alliances and various forms of marketing joint ventures are lost to the rival, or others switch over to you.
6. The defender is discouraged, and it is, to some extent, neutralized.

The causes of loss in strength include:

1. You aggressively challenge the rival's strong areas and overspend unrecoverable resources. Whereas the defender may have access to unused resources or receives additional budgetary commitments from its management.
2. The moment you enter the competitor's market, the nature of operations changes. It becomes contested through claims and counterclaims, price wars, increasing promotional expenditures, and similar challenges that were totally unexpected.
3. You move away from your primary supply chain, while the defender tightens his ties to his own. This causes delays in establishing new relationships.
4. The danger threatening the rival brings possible aid from such areas as venture capitalists, offers for joint ventures, and opportunities for mergers or to be acquired. The latter two from organizations that seek an opportunity to gain a quick entry into a market.
5. The competitor, facing heightened danger, makes a spirited effort to fight back, whereas your efforts slacken.

All of these gains and losses in strengths may co-exist. The essential point is that they show the full range of alternating effects embedded in a campaign. The outcome depends on their power to arouse either side to greater efforts, or to cause one side to retreat.

There are other essential points to consider. For instance, it is not possible in every contested market for the winner to completely overthrow a competitor. That is, even a partial victory has a culminating point, which normally results from the superiority of one side having achieved an optimum mix of physical and psychological strengths.

As a competitive campaign unfolds, organizations are constantly faced with some factors that increase their strength and with others that reduce it, as listed above. The question, therefore, is one of superiority.

Every reduction in strength on one side can be considered an increase for the other. That results in a need to rebalance the ratio of strengths to weaknesses, which at times can occur through internal changes.

For instance, IBM, also referred to in Chapter 7, during 2014, was in the middle of implementing bold new approaches to turning around the massive organization. In so doing, CEO Ginni Rometty had to calculate an ending point by eliminating its lower-margin businesses as she shifted resources to strengthen its move into cloud services and commercializing Watson, with its supercomputer capabilities. The moves were determined within the overall framework of IBM's turnaround plan.

That is where leadership is vital. If the leader is not easily visible and available, a serious handicap arises from the loss of time involved in the decision-making process. Even the expanded authority given to a local manager will not suffice to meet every contingency that may arise in his or her sphere of responsibility.

Absence of leadership is especially troublesome when opportunities are missed and the best chance to recover a threatened situation is wasted, whereas, boldness and the willingness to take rapid assertive action is needed.

In any event, an ill-considered continuation of an offensive campaign may drain the strength and resources with which you began, or reduce the effect of what you gained from the initial victory.

If this effect is possible, why logically persist in pursuing the offensive? Would you still call this a victory? Would you not do better to end before you begin to lose the upper hand?

The answer is that superior strength is not the end but only the means. The ending point is either to neutralize the competitor or to possess some of the rival's market. In either case you improve your general prospects in the total effort.

Therefore, to reinforce the essential point, be watchful of overshooting the target. Otherwise, instead of gaining new advantages, you could harm yourself.

Expressed in still another way: If you were to go beyond the end-point guidelines identified above, it would not merely be a useless effort, it could not add to success.

Consequently, your obligation, and that of every line manager is to stay as close as possible to an evolving campaign and determine the critical ending point when an offensive turns into a defense. That point also occurs when the rival's capabilities are neutralized.

For many managers, however, as long as an offensive campaign progresses, there must still be some superiority with which to keep going. Further, since defense would start when the advance ends, there is not really much danger of suddenly becoming the weaker side.

The risk of a setback often does not reach its peak until the offensive move has lost its momentum and it turns into a defense. It also should be obvious that all these factors will not be found in equal strength in every situation. Therefore, one defense is not exactly like another, nor will every defense enjoy the same degree of superiority over attack.

Meanwhile, should your offensive action continue, you sweep across the threshold of equilibrium (the ending point) without knowing it. It is possible that in your state of mind, reinforced by the psychological benefits peculiar to offensive movements, and in spite of your diminishing resources, you may find it easier to continue rather than end the campaign, a condition usually reinforced by the zeal of the staff to keep moving.

That is how an attack can reach the point that results in exhausting the budget. Also an alert competitor may recognize this error and exploit it to the fullest. That is why calculating the end point correctly when planning the campaign is essential if you are to secure your turnaround plan and preserve your resources.

In reviewing the whole array of factors you must weigh before making your decision, you must realize that numerous wrong turns and mistakes are possible from so many different directions. And, if the range, confusion, and complexity of the issues are not enough to overwhelm you, the risks and responsibilities may influence you. This is why a great many leaders will prefer to stop short of their objective rather than risk approaching it too closely.

Then, there are those leaders with high courage and an enterprising spirit who will go beyond the reasonable end point and reach out for the possibility of achieving a resounding success, and they consciously do so knowing the risks and responsibilities.

They are the great leaders that create industries, organizations, and fortunes for themselves and others. In the end, however, the odds favor both types of leaders, if they acknowledge the value of determining the end point of a campaign and use the guidelines discussed here.

VALUING SURPRISE AND SPEED

In summarizing the merits of surprise and speed, value them highly and make them an integral part of your turnaround plan. Any notion that speed means reckless or impulsive moves totally misses the advantages. Rather, it means making research-driven and prudent estimates prior to taking action.

Once a decision is made, procrastination creates a formidable barrier to a successful outcome. Your aim is to move forward immediately to neutralize your competitor's edge.

From a total company viewpoint, and operating within a global setting, speed impacts a number of managerial, organizational, and competitive issues. For instance, it affects such areas as:

- How you organize your company's or business unit's pecking order so that your decisions flow smoothly without getting stuck in a maze of managerial layers, which often results in distortions and misinterpretations of what was originally intended.
- How fast you react to a global competitor oceans away, where its clear-cut aim is to feed off your customers and erode your market position in your local marketplace.
- How confident you are in your turnaround plan, so that preplanned strategies and tactics are set in motion in a timely manner, rather than react with hasty, fit-and-start movements.
- How and when you launch a new product to gain a competitive edge and secure a favorable share-of-mind position among early adopters.
- How rapidly you harness social media and integrate it into your total marketing program, so that the technology impacts favorably on customer solutions.
- How fast you adopt new systems to foster virtual communications within your organization and along the entire supply chain.

"There is no instance of a country having benefited from prolonged warfare," declared Sun Tzu.

To paraphrase such sage advice: There are few examples where an organization has benefited from prolonged, drawn-out campaigns. Tardiness, sluggishness would unquestionably affect many of the above internal functions of your organization and influence your ability to maintain the morale of your employees.

Further, lethargy can be traced to an ingrained cultural trait of the organization, which (if it exists in your firm) reflects on your managerial style and, in turn, could inhibit your ability to lead under real-time market and competitive conditions. Therefore, you can improve your chances for securing a turnaround by acting with speed and surprise.

Section IV

Secure Competitive Advantage and Preserve Success

- Use a Pretest to Evaluate Your Turnaround Plan for Competitive Advantage

10

Use a Pretest to Evaluate Your Turnaround Plan for Competitive Advantage

Chapter Objective

Be able to

- use a set of nine guidelines to pinpoint the key factors for pretesting the effectiveness of the turnaround plan.

INTRODUCTION

Developing a turnaround plan is a mental process, joined to a physical, action-oriented series of strategies. It means acquiring a positive mindset that conceives of an idea, expresses it as a vision or strategic direction, and shapes it into tangible outcomes. From there, the mind conceptualizes the results into a useable form that you can use to inspire those you manage. Only then does it begin to take discernable physical shape.

Presumably at this point you have internalized the broad concepts that go into developing your turnaround plan. Perhaps, too, you solidified the ideas and the first steps are underway to develop the plan.

From the point of view of this chapter, then, the assumption is that you have a draft of a plan that you initiated as a result of a competitive crisis or an existing condition that alerted you to an impending problem. Before sealing it, however, some procedure is needed to pretest its completeness. The following nine guidelines make up the pretest.

Conditions Triggering a Turnaround

The central idea behind most turnarounds is that some preexisting conditions precipitated the need, and possibly the urgency, to undertake a turnaround. These circumstances may go beyond surface issues of, for instance, a new competitor entering the market at lower prices.

The actual reasons may require you to drill down for causes that were endemic to the problem and could not be fixed by any of the generally accepted practices, such as adjusting budgets, introducing austerity programs, or shelving projects and new product introductions.

Assuming the problems originated from competitive actions, the following questions provide a checklist to pretest key areas of your turnaround plan:

- Do we have a warning system in the event a competitor moves into our market with an aggressive strategy or a disruptive technology that would immediately give it a competitive advantage and place us into free fall?
- Do we have a contingency plan ready to counteract sudden moves? Or are we still inner-focused and complacent, believing we have the total loyalty of our customers, without regard that there is an active competitive environment with rival companies looking for growth opportunities? (See Chapter 7, Levi Strauss case.)
- What is likely to be the immediate response by our management at the time of an aggressive move by a competitor: actively engage the competitor, go into a protective mode, or retreat by viewing the situation as untenable?
- Is our organization structured so that there is a flow of meaningful intelligence from the field to the decision-making managers who can take quick action?
- How would our personnel likely react to the news of an attack on our main line of business? How effective is our leadership in monitoring their morale and behavior?
- To what extent does our leadership understand that if surprise does occur, it is not caused by some random occurrence or isolated event?
- Are we able to detect the type of competitive campaign that we are liable to face, so that we can decide on the type of strategy to employ?

The Organization

The organization is a living entity in which individuals work. Made up of the physical and psychological dimensions, it is managed by a leadership, directed by a strategic business plan (if it exists), and operated by trained and motivated personnel.

It is propelled by creative strategies supported by competitive intelligence that aims to coordinate the means (money, human resources, and materials) to achieve the ends (profit, customer satisfaction, and company growth) as defined by the company's policies, its strategic direction, and objectives. More succinctly, the organization's purpose, and the object of business, according to the renowned Peter Drucker is "to create a customer."

Centered within the organization's inner core is its culture made up of its values, visions, symbols, beliefs, and history. Culture contains the organization's distinctive DNA that sets it apart from others. In effect, it shapes the behavior of personnel and reflects in their feelings and actions.

The following questions provide a pretest about your organization:

1. Can our organization be characterized as operating with speed and efficiency? Or is it bogged down with organizational layers that complicate communications and decision making?
2. To what extent are our personnel trained and motivated to pursue a turnaround?
3. Do we have an internal system for the exchange of information among individuals in different functions of our organization, so that disruptive technologies and events can be identified in a timely fashion?
4. Is there a mechanism for the staff to cooperate with each other in developing plans, such as would be the case with a cross-functional strategy team?
5. Is our organization's culture in tune with today's operating environment, and, in the framework of a long-term turnaround plan, is it properly aligned with the overall direction of the plan?
6. At what level is our employees' morale and is it conducive to sustaining an energetic effort for a turnaround?
7. Are there pockets of negative behavior that would adversely affect developing and implementing a turnaround plan?

The Turnaround Strategy Plan

There are numerous reasons for developing a plan. One obvious reason, and the purpose of this book, is to provide a process that results in a course of action to develop a turnaround for an organization facing, or is about to face, a problem derived from a competitive situation. Another functional reason is that a plan provides a convenient "housing" for all the history, current conditions, and desired outcomes to be clearly stated. Still another purpose is that a strategic business plan furnishes the platform for strategic thinking.

Then, in further support, one well-known management consulting firm conducted a survey of large and midlevel organizations that revealed the following:

> Two thirds of rapid-growth firms have written business plans. Firms with written plans grow faster, achieve a higher proportion of revenues from new products and services, and enable CEOs to manage more critical business functions than those firms whose plans are unwritten. Additionally, growth firms with a written business plan have increased their revenues 69 percent faster over the past five years than those without written plans.

Further, three other dimensions can be added to the above reasons for a plan.

First, a written plan defines the organization's strategic direction, lists objectives and strategies, and provides details about products and services. Implicit in the physical makeup of the plan are the numerous details about tactics to reach customers in geographically and culturally dispersed markets, using the seemingly limitless potential of the Internet, Big Data, and the vast potential of such supercomputers as IBM's Watson (see Chapter 7).

Second, planning is a mental process, and the plan permits the juncture where experience, skill, and insight converge to envision the future. It is the point at which to deploy people, material, and financial resources for maximum impact, as well as to assign levels of authority and responsibility to trained individuals who can skillfully implement the plan.

Third, the plan adds credibility and purpose to leadership, which can energize the morale of the rank and file to push forward in an increasingly competitive environment. A well-conceived plan that is understood and accepted by the staff is often the singular factor in effective leadership.

Therefore, the plan binds those three dimensions together into a cohesive and managerial whole, to make sense of the organizational and marketplace arenas, and to create order out of what can easily deteriorate into disorder.

The following questions provide a checklist as a pretest of your turnaround plan:

1. Do we now have an internal working environment and a process that permits formal planning to take place, as opposed to one that relies primarily on an annual budget-submitting ritual?
2. In the interest of making planning participative, are individuals trained in the process, and have we identified those who need such training?
3. As we continue thinking and deliberating over campaigns and strategies, do we still give ample time and attention to the preexisting conditions that triggered the need for a turnaround and the remedial actions that have been employed?
4. Are we focused on the key issues related to implementing the turnaround plan, such as overcoming the causes of friction, maintaining a customer-driven orientation, and developing strategies to neutralize the competitor?
5. While time is given to systems, gathering data, and overcoming existing problems, are we continuing to give sufficient time to envisioning the future, thinking about the kind of business we will be conducting in the next five years, profiling the types of customers we will be serving, weighing new functions we may be performing based on customer needs, looking at the new-wave technologies on the horizon, and anticipating the new competitors we will be facing?
6. Are we sufficiently aware that the objectives and strategies that we develop must feed off the strategic direction of the plan, i.e., if the strategic direction is too narrow in scope the objectives and strategies will follow that span; if, too expansive, our objectives and strategies may be too far a reach based on resources, the staff's capabilities, and the existing culture of our organization? Do we have a proper balance between the two dimensions?
7. Are we fully aware that a major component of the plan must focus on strategies, which are the action elements that set the plan in motion? ("All plans must deteriorate into action," declared Peter Drucker.)

Leadership

Creating a turnaround plan and implementing it with success involves many factors: people, resources, resolve, and commitment. Then, there is one compelling influence that molds all the factors into a potent force: leadership.

To function as a leader means providing purpose and direction to improve the long-term viability of the organization. Therefore, anyone responsible for supervising people or accomplishing its organizational mission by managing people and resources is a leader.

Taking it a step farther, any individual who influences and motivates people to action or affects their thinking and decision making is a leader. Leadership is not only a function of *position,* it is also a function of an individual's *role* in the organization.

In a turnaround situation, there is an additional level of sensitivity needed to pull the organization out of decline, reverse direction, and make an upward sweep. It takes a form of leadership that is highly personalized, yet flexible. Anything else will come across to your personnel as artificial and insincere. This is especially important if you expect them to support the overall organization's strategic direction and objectives.

Also, personalized and flexible mean that you shouldn't rely only on one leadership style. Otherwise, you suffer the effects of being overly rigid and you will likely experience difficulty operating in situations where a single approach simply doesn't work. Meaning that some projects are complex and require different management skills at each stage of development.

For instance, projects in the early stages of development, where creative oversight on performance and quality dominate, require a far different leadership style from that of pumping up a sales force when launching a new product.

Similarly, products at various stages of their life cycles—introduction, growth, maturity, decline, and phase-out—involve different leadership methods to correspond to the varying market and competitive conditions at each stage.

For those reasons, there is no single leadership style for all occasions. Therefore, your style should match your organization's overall objectives. Be certain, too, that it conforms to the individual tasks to be performed by role and function.

The following guideline questions provide a pretest of leadership:

1. Is our current leadership, at all levels, capable of tracking causes related to low morale among the various groups in the organization, and is it able to elevate the feelings of confidence, self-esteem, and optimism among those individuals?
2. Do our executives and midlevel managers display the courage to make bold moves?
3. Beyond intellect and courage, do our leaders exhibit the determination to see the plan through to a successful end?
4. Do our leaders possess a presence of mind to think quickly in the face of the inevitable barriers that will be present during the turnaround?
5. Is strength of mind shown in the leaders' abilities to retain a sense of calmness and self-control during the stressful periods?
6. Do the leaders demonstrate strength of character and balanced temperaments with the ability to remain calm and firm, and not be rattled by negative events?
7. Can our executives merge the diversity of information about our organization's culture, employee morale, available resources, and market intelligence with the plan's strategic direction, objectives, and strategies and still keep a proper perspective of the total situation?

The Competitive Campaign

The competitive campaign has a structure and is impacted by various forces.

First, there is the pervasive effect of *friction* that exists as internal and external hindrances. It is found among the functional groups where there is often an ongoing adversarial relationship, such as with manufacturing versus marketing or finance versus product development. Then, there is the friction resulting from physical and mental fatigue from an overworked staff.

Friction also is associated in the marketplace where the firm's product may not be compatible with the customers' operating systems, or where market intelligence is sketchy and inaccurate. Then, there is the natural tendency for individuals to assume the worst about news from individuals who tend to exaggerate or misinterpret events. Add to that the innumerable small things that can complicate seemingly minor tasks.

Second is the effect that *neutralizing* the competitor's ability to interfere with your efforts has on achieving your objectives. That means taking actions that reduce the competitor's strength relative to your own.

The aim, then, is to locate and exploit the rival's weaknesses and vulnerabilities. If successful, a neutralized competitor would be at a competitive disadvantage and discontinue the campaign without incurring excessive losses.

Third is the *duration* of a campaign. Most campaigns are rarely decided in one event. Consequently, a campaign shouldn't be given up, except where the objectives have changed, or where there is predetermined ending point based on a variety of factors, such as resources, the morale of the staff, better opportunities elsewhere, or continuing the effort until it turns out to be a fruitless effort.

Fourth is the issue of *defense* versus *offense*. That means looking at the advantages and disadvantages of defending what already exists versus exploiting new markets and opportunities. One central issue in favor of defense is that retaining existing customers is far more profitable and easier than paying heavily to acquire new customers. It is also more advantageous to hold a well-entrenched market position than to occupy a new one.

Whereas, on the surface, the defense is advantageous to attack, relying totally on that approach is contrary to building for growth. The optimum situation is to secure your position defensively, then move to the offensive. Also, where offensive moves are necessary for growth, there has to be total awareness of the practical realities of budgets being exhausted and management's attention being shifted elsewhere.

Fifth is the use of *reserves*, which is a strategy decision based on the following three considerations: (1) prolong a campaign where the rival doesn't have the resources to hold out against you for an extended period of time; (2) counter any unexpected threat from a competitor; (3) exploit a market breakthrough with a new product, or take advantage of an unforeseen weakness in the rival's defenses.

Altogether, the essential purpose of reserves is that additional resources are available at a decisive stage in a campaign where a full concentration of effort is needed to assure a successful outcome.

Use the following guidelines questions as a pretest about the structure and characteristics of your campaign:

1. Are we aware of the numerous causes of friction that can bring our campaign to a halt, and do we know how to reduce their damaging effects?
2. In particular, are we mindful of the possible negative effects of friction among various groups and key individuals within our firm that can consume valuable time and create a state of nonaction?
3. Do we understand that a turnaround consists of numerous individual campaigns; that a failure in one should be regarded as merely a temporary condition for which success is possible in subsequent campaigns?
4. Have we fully internalized the idea that the important goal of a competitive campaign is to neutralize or cancel out the effectiveness of the rival's capabilities?
5. Are we alert to the advantages and disadvantages of defense versus offense when determining campaign strategies?
6. Do we understand the primary purposes of maintaining and using reserves?
7. Are we aware that what we set aside as reserves should be compared with those available to the competitor (if known); if they exceed ours, it may be necessary to reexamine our campaign strategies?

Bold Action

Boldness is a preferred course of action. Provided, that is, it is supported by sound business plans, reliable estimates of competitive and marketplace conditions, ongoing competitive intelligence, and a solid commitment to your company's turnaround objectives.

These supporting activities prevent you from taking mindless risks by entering markets that are not defensible and that are beyond your company's financial resources, supply chain strengths, and employees' capabilities.

Consequently, if those supports are in place, your boldness is justified. And you vastly improve your chances of extending your horizons, with the strong likelihood that you will end up with profitable outcomes.

There is, however, the highly realistic situation in which you face a market circumstance that is time sensitive and requires you to take some action or forfeit a potentially lucrative opportunity to a competitor. That means having to live with gaps in competitive intelligence. In that case, you have little choice but to move ahead with the information on hand, relying on your experience, self-confidence, flexibility, and intuition to face up to the market condition as it unfolds.

Bold action also has a powerful psychological impact on leadership in that it elevates such traits as confidence, self-assuredness, and initiative. Whereas, any signs of excessive caution displayed by a leader often shows up as a loss of motivation, energy, and momentum. There is also the negative effect on employee morale where a heightened level of energy and resolve are needed to maintain the push.

Boldness, however, in no way implies careless and impetuous movements. It does mean that once the situation has been carefully assessed, and it is backed up with market intelligence and calculations, then bold action is required to make things happen. Other factors being somewhat equal, when comparing boldness to caution, boldness is the preferred course of action. Excessive caution is more often than not worse than too much courage.

The following questions provide a checklist for your self-test:

1. In a declining situation, does our leadership exhibit the kind of boldness to reverse any further decline?
2. Are there signs of an excessive show of caution on the part of senior management?
3. If bold action is shown, should it be considered an ingrained cultural trait, or is it localized to one or two individuals within the organization?
4. Whereas boldness requires accurate competitor intelligence, is there a reliable procedure in place for assessing competitive operating patterns for telltale signs of opportunities or threats?
5. Are there forthcoming marketplace situations threatening, especially from rivals, that would suggest taking a more cautious approach?
6. Are we using any of the classic management tools for determining the bold versus cautious approach?
7. Based on the responses to the above questions, is it necessary to reconsider the turnaround objectives we set?

Concentration versus Dispersal Strategy

Concentrating your effort at precisely defined customer segments is in sharp contrast to spreading resources over numerous areas in a seemingly play-it-safe mode. This classic principle is based on the strong evidence that concentrating resources where you can gain superiority in a few decisive areas will provide the best chances for winning.

Succeeding with concentration incorporates a three-step process: (1) define customer segments and evaluate competitors' vulnerable areas; (2) use market data to pinpoint a segment for initial entry, followed by a systematic rollout into additional market segments; and (3) customize products and services to those segments.

Concentration is particularly appropriate when maneuvering against a market leader. You have the opportunity to be highly selective and indirectly take a segment-by-segment approach to avoid its strong points. That is, the application is feasible where that larger competitor is following a concentration strategy as well.

That means, your rival may leave some segments unattended or marginally served. These could be your point of entry to secure a strong position. You, thereby, can achieve an advantageous competitive position in those market segments, as long as they correlate with your core capabilities and correspond with the objectives of your turnaround plan. The essential point, then, is that concentration permits you to deliberately focus your resources at areas that represent pivotal points for further maneuver.

Use the following guideline questions to assist in your self-test:

1. Are we aware of which segments and market niches represent decisive points and where our choices are supported by market data?
2. Do we use some form of comparative analysis to pinpoint which areas represent the decisive points of vulnerability among competitors?
3. To what extent are our senior executives and midlevel managers capable of accurately singling out decisive points and shaping cohesive turnaround strategies?
4. Within the framework of marketing and logistics to reach the ultimate user of our products and service, what media and methods would justify our concentrated efforts? Or, if some measure of dispersal is prudent, what mix would make up the optimum balance?
5. How flexible are we in moving from underperforming segments and shifting resources to concentrate in more lucrative ones?

6. To what extent have we tied in research and product development to fit the requirements of a concentration strategy?
7. Do we need to make any organizational changes to activate a concentration strategy, such as coordinating activities with marketing, communications, distribution, and specialized services?

Indirect versus Direct Strategy

The indirect strategy is consistently one of the dominant and enduring principles of competitive strategy. It is a vital component in developing and implementing a turnaround plan. It is particularly essential if strained by limited resources and pressured to curtail losses.

On the other hand, a direct strategy tends to go head-to-head against a competitor with obvious moves. This is especially the case where there are few areas of differentiation in product benefits, quality of service, perceived price, and the like. The end result is battling it out in the marketplace in a manner that results in resource-draining price wars.

The indirect strategy operates on three dimensions.

First, the strategy is anchored to a line of action whereby you apply your strength against a competitor's weakness. The essence of the move is to maneuver so that your rival will not or simply lacks the capability to challenge your efforts.

Second, concurrent with activating indirect moves against a competitor, you focus your attention on serving customers' needs or resolving their problems in a manner that outperforms your competitors' methods.

Third, you aim to achieve a psychological advantage by creating an unbalancing effect in the mind of the rival manager. That is, by means of distractions and sudden moves, you make it appear that you are launching your effort directly at the competitor's strengths. Whereas your true purpose is to target his vulnerabilities.

As one ancient sage, Sun Tzu, commented, "The direct method may be used for joining battle, but indirect methods will be needed in order to secure victory."

Additionally, the psychological effects of the indirect strategy seeks to disorient the competing manager, causing him or her to waste time, effort, and resources in the wrong direction, or, expressed differently, make costly and irreversible mistakes.

All three applications serve the strategic purpose of reducing any resistance leveled against your efforts. In turn, you can then utilize the full power of your resources without wasting them on strength-draining actions. Therefore, look at the indirect strategy as an encounter of manager against opposing manager; your experience and skill pitted against those of your opponent.

The following questions provide guidelines for your self-test:

1. Considering that developing indirect strategies results from thinking strategically and creatively, to what extent do the maneuvers we are considering link to our strategic turnaround plan?
2. Do our managers understand that indirect strategies aim to outthink, outmaneuver, and outperform existing and emerging competitors, and that means implementing them with little-to-no direct confrontation?
3. A subset of thinking strategically is acting tactically; how, then, are field managers and sales personnel tuned in to thinking of indirect approaches?
4. Do our managers actively probe for unserved market niches where there is minimal resistance from competitors, and where possibilities exist to establish a foothold and then indirectly expand into a mainstream market?
5. To what extent do evolving technologies permit maneuvering around an overcrowded supply chain and reaching straight out to the end user?
6. How effective are managers in using the marketing mix as a source for developing indirect strategies?[*]
7. Indirect strategies combine the physical with the psychological to create an unbalancing effect on the rival manager; how effective are we in creating such an outcome?

Surprise and Speed

It is a generally accepted truism at various levels of management that any campaign prolonged for an extended period of time loses its effectiveness; it is as if a spring has uncoiled and lost its power. Worse, yet, resources are often consumed beyond their budgeted limits, and morale suffers from long periods of excessive stress.

---

[*] See Chapter 1, Table 1.1, which includes the traditional marketing mix, plus other areas as sources for developing indirect strategies.

Many of the obstacles to acting swiftly and decisively originate from such common organizational forms of friction as extended deliberations that consume valuable time, cumbersome committees that drag along, and the overall pattern of putting off making timely decisions.

Damaging, drawn-out efforts affect personnel as they become bored and lose their edge. Then, there are the gaps created through lack of action that give competitors extra time to create barriers that would blunt your next efforts. As a result, opportunities simply evaporate or the competitor acts and gains the advantage.

On the positive side, speed of action adds vitality to a company's operations and acts as a catalyst for growth. As a major factor toward competitiveness, it impacts virtually every part of the organization.

A parallel concept associated with speed is surprise. As indicated in developing indirect strategies, the aim of surprise is to create an unbalancing effect and cause the rival manager to make faulty decisions. The aim, again, is to prevent a direct confrontation and assure success of your maneuver.

The following questions provide guidelines to serve as a self-test:

1. Have we fully internalized the idea that even minor delays can result in a loss of momentum and could signal a vigilant competitor to move in and fill the gap?
2. Do we understand that a strategy that integrates speed with technology is in a prime position to secure a competitive lead, which makes attempts at catch-up difficult to achieve?
3. To what extent do our managers fully comprehend that losses in market share and competitive position often result from prolonged delays, especially when dealing with time-sensitive market conditions and quick-moving competitors?
4. Are we aware that speed and surprise require a committed and spirited group of individuals who can react quickly to unfolding events; whereas a group that exhibits a general malaise will result in missed opportunities?
5. Do we fully comprehend that any excessive loss of time in responding to market developments with new products increases the chance of existing products reaching maturity or a commodity level?
6. Are we alert to where organizational obstacles and other forms of friction still exist that can prolong deliberation, foster procrastination, and delay decisions?

7. Do we recognize that surprise with its powerful psychological effect should be purposely made part of every competitive campaign?

A FINAL WORD

The underlying assumption of this book is that if a crisis existed in a firm, your firm, any firm, that resulted in management calling for a turnaround plan, it was likely triggered by external market conditions. More specifically, they were initiated by competitive actions that were intense and totally unexpected.

Yet those conditions and actions were unforeseen. Why? Were executives unaware of what was going on in the marketplace, especially among competitors? Did they completely miss a new technology cycle? Or were they overly late in noticing changing buying practices?

Did the managers think they had a lock on the market and couldn't be dislodged? Was it a complacent, inner-focused corporate culture that relied on its history and outdated values to sustain it during the good and bad times? Or were there deeper reasons that are yet to surface?

Are only the smaller, more vulnerable companies victims of such conditions, or are major organizations vulnerable as well? Consider the market giants that faced one or more of these questions and had to come up with answers and remedies. For starters, think Kodak, Intel, Staples, Hewlett-Packard, General Motors, Panasonic, Dell, J. C. Penney, McDonald's, Best Buy, IBM, BlackBerry, and Levi Strauss.

The essential points behind these questions include the following:

First, any thought of developing an effective turnaround plan should be based on the premise that faulty conditions preexisted within the organization that caused marketplace events to reach crisis levels. Although a certain level of chance and luck exists in every situation, nevertheless where events hit with enough force that can permanently damage the firm, there is something intrinsically wrong in the company far beyond a run of bad luck.

Second, before any thought of developing a turnaround plan gets underway, the root causes of the problem need to be identified and steps taken to remedy the situation, or else the problem could linger and degrade into an even more critical state. (These conditions and remedies are detailed in Chapters 1 and 2.)

Third, attempting to solve the problem by taking sudden and sometimes impulsive action of immediately laying off personnel, closing facilities, and chopping projects may be premature and serve only as a temporary fix. Doing so is not likely to cure the deep-rooted problems that initially created the need to turn around the organization.

Next, as you look at the structure of the turnaround plan (Chapter 3), it appears somewhat like a format for a strategic plan. To some extent it is. The rationale is that this type of planning forces you to look out in time and reimagine what your organization would look like in, say, five years; envision what audiences you will serve; and conceive of the products and services you will offer.

Then, once the turnaround plan is complete and put into action, a follow-on task is to preserve success. That means preparing your mind for the challenge. If your mind gives in to complacency when you have a purposeful strategic direction, objectives, and strategies, then you will have proved nothing.

If, however, you adopt the mindset that the outcome of a single campaign is never to be regarded as final; that is, even if it suffers negative outcomes, it should be viewed as merely a transient event. Remedies may be found in a variety of possible solutions as you move forward.

It is at that pivotal point where you have to take on the transforming effects of courage. It is where you listen to your inner voice of intuition and give in to the power of determination and perseverance. All such positive thinking can make the difference between success and failure. Thus, presence of mind, strength of mind, and strength of character become the true deciders of a turnaround (see Chapter 4).

What follows is that your bearing, thinking, and attitude must inspire and motivate others through effective leadership. The *others* are those individuals, your staff, who are also capable of erratic behavior that can alternate between laid-back or spirited, passive or active, indifferent or enthused.

In all circumstances, they are the ones you must reach and elevate through effective communications and ongoing training. They need to be viewed as valuable assets that can contribute added value to overall strategy discussions or to a specific idea—ideally as part of a cross-functional strategy team (see Chapters 2 and 4).

As you reach into the future and attempt to project your thinking in the light of disruptive technologies, new and existing competitors, and changing patterns of customer behavior, a poststrategy is needed to sustain the momentum and prevent the same threatening conditions to surface again (the poststrategy process is described in Chapter 3).

Finally, a vital factor that forms the underpinnings of your turnaround plan is strategy. It is the action component that achieves the desired results as stated in your objectives.

Good luck on your turnaround.

Index

A

acceptance *vs.* anxiety, 44
actions and activities
    acting like aggressive competitor, 35
    competitive crisis *vs.* random occurrence, 5–6
    indirect strategies, *vs.* direct strategy, 172
    some better than none, 34
    zones of, 55
active participation, 44–45
Adesto Technologies, 129–130
advertising, tactical, 178–179
agent utilization, *see also* Stealth activities
    double agents, 161–162
    expendable agents, 162
    general agents, 160
    inside agents, 160–161
    living agents, 162–163
    overview, 159–160
air conditioner manufacturer example, 14
alertness, 50
Amazon
    blaming, 79
    conflicts, not isolated events, 19
    dispersal strategy, 158
    friction reduction, 73
    Google's internal conditions, 105
    offense, 111–112
    speed, 193
    strategic thinking, 169
    strength of character, 91
ambition, 15, 88
American Airlines, 197
Android, 18, 173
anthropological approach, 15
anxiety *vs.* acceptance, 44
Apple
    dispersal strategy, 157, 158
    Google's internal conditions, 105
    joint ventures, 150
    neutralizing competitors, 17–18
    Panasonic comparison, 4
    point of friction, 100
    regulatory issues and industry trends, 151
    speed, 193
    tactical maneuvering, 173
Apple TV, 91
application failure, 70
Arizona utility company, 182
arrogance, 142
Arthur D. Little Matrix, 136–138
artifacts, culture, 38
automated radioimmunoassay, 137
awards, *see* Rewards and recognition
awareness, 15, 65

B

balance, 101–102, *see also* Unbalancing competitors
Barnes & Noble, 185
BASF, 7
BCG growth-share matrix, 132–134
Begemann, Brett D., 30, *see also* Monsanto
behaviors, *see also* Psychological dimension
    conflict visibility, 46
    consequences, 51
    difficulty interpreting, 105
    inaccuracy and unknowns, 18–19
    inconsistent *vs.* flexibility, 90
    leadership, 199
    response to negative, 51–52
    stubbornness, 90–91
Bergh, Chip, 142–143
Best Buy
    company size, 223
    competitive campaigns, 7
    defensive campaigns, 8–9, 23
    physical dimension, 28
    psychological dimension, 29
    reenergizing demoralized personnel, 14–15

reverse declining sales campaigns, 14–15
bet-the-company culture, 38
Bezos, Jeff, 91, 120, *see also* Amazon
BlackBerry, 17–18, 223
blood collection system, 138
Bloomingdale's, 181
BMW, 7, 197
bold action *vs.* cautious restraint
    Arthur D. Little Matrix, 136–138
    BCG growth-share matrix, 132–134
    boldness, 123–127
    caution, 128–131
    decision making tools, 131–140
    decisive points identification, 128
    employee mindsets, 33–34
    General Electric Business Screen, 134–136
    management by objectives, 138–139
    offensive actions, 33–34
    ongoing competitive activity, 20
    overview, *xvi*, 119–123
    pretest, plan evaluation, 217–218
    question checklists, 218
    Six Sigma, 139–140
Bonaparte, Napoleon, 86, 183, 191
book stores, *see* Amazon; Half Price Books
Borders, 184
branding, 35, 197–198
Brin, Sergey, 120, 123
broader interpretation, 59–60
budget, *see* Reserves, use of; Resources
business strength, 135–136

C

calmness, 89, 91
campaigns, *see* Competitive campaigns
Campbell's, 24
Canon, 8, 110
cash cows, 133
Caterpillar, Inc., 103
cautious restraint, *see* Bold action *vs.* cautious restraint
celebrities, 17
central markets, 95
chain of command reduction, 29
challenging markets, 95–96

Chambers, John, 191, *see also* Cisco Systems
chance
    decision impact, 83
    elements of, 25–26, 54
    postcampaign strategies, 65
changes
    awareness, 58
    competitive campaign, 99–115
    leadership techniques, 79–98
    overview, *xvi*
character, leadership, 90–92
characteristics, 99–115
checklists
    bold action, 218
    competitive campaign, 217
    competitive crisis *vs.* random occurrence, 5–6
    concentration *vs.* dispersal strategy, 219–220
    endpoint of campaign, 113–114
    indirect *vs.* direct strategy, 221
    leadership, 215
    organizational conditions, 211
    strategy plan, 213
    surprise and speed, 222–223
    turnaround plan, key areas, 210
Chef Collection, 47
Chevrolet, 197
Chipotle, 158
Cisco Systems, 169–170, 191
City Winery, 191
Clinton, Hillary, 17
cloud computing system example, 73
college classrooms, 160
Comcast, 35
communications
    alertness, market conditions, 50
    employee input, 32–35
    factors, 149
    indirect *vs.* direct strategy, 172–173
    ineffectual leadership, 69
    organizational culture, 31
    strong internal, 36
company newsletters, 161
comparative analyses, 72
competitive advantage, *see* Pretest for evaluation
competitive campaigns

Index • 229

characteristics, 99–115
components, 104–115
conducting, 107–109
defense, 109–111
duration, 106–107
ending point, 72
expansion into additional markets, 10, 12
follow-up, 112–114
increasing costs for rivals, 15
limited-term *vs.* long-term, 14
new markets, long-term objectives, 13
obligatory commitments, joint-venture agreements, 10
offense, 109–112
opportunities, 10, 11
overview, *xvi,* 7, 99–104
preemptive, 10
pretest, plan evaluation, 215–217
prolonged, 205
question checklists, 217
reclaiming former market, 8
reenergizing demoralized personnel, 14–15
reserves, use of, 114–115
retaining share of market, key region, 8–9
reversing declining sales, 14–15
slow moving, 191
solidify position, 13
structure, 99–115
upward pressure, ambitious objectives, 15
weakening defender's resistance, 13–14
competitive conditions, 50
competitive crisis *vs.* random occurrence, 5–6, 19–20, 224
competitor/competitive intelligence
boldness, 124–126
campaign standstill, 22
concentration *vs.* dispersal strategy, 159–163
inadequate, 68
overestimation of strength, 24
priority, 23–24
Sun Tzu viewpoint, 21
surprise, 187–191
vulnerabilities, 141
competitors

concentration strategy implementation, 150
employee input, 35
expectations for plan, 47–49
fear, 183–184
inadequate intelligence, 68
physical/psychological characteristics, 17–19
primary strategies, 71–72
stress, 182–183
unbalancing, 181–185
competitors, neutralizing
Best Buy example, 9
boldness, 124
competitive campaign, 216
competitor's strategies, 47–49
overview, 104–105
physical and psychological characteristics, 17–19, 54
primary strategies, 71–72
success, 108
complacency
conflicts, not isolated events, 19
Levi Strauss & Co., 142
ongoing competitive activity, 20
postcampaign strategies, 66
components, competitive campaigns, 104–115
concentration, decisive point, 71
concentration *vs.* dispersal strategy, *see also* Resources
agent utilization, 159–163
competitive intelligence, 159–163
competitors, 150
concentration strategy implementation, 144–156
consumers, 144–145
dispersal strategy implementation, 156–159
distribution method changes, 149–150
distributor changes, transitions, 147–149
double agents, 161–162
expendable agents, 162
general agents, 160
industry trends, 151–152
inside agents, 160–161
intermediaries, 146–150
leadership, 152–153

living agents, 162–163
market coverage, 147
market research, 152–153
new product introduction, 146–147
organizational issues, 153–154
overview, xvi, 141–144
planning, 153–154
pretest, plan evaluation, 219–220
question checklists, 219–220
regulatory issues, 151–152
SWOT analysis, 163–165
utilization guidelines, 155–156
conditions, *see* Organizational culture and conditions
conflicts and confrontations
    circumventing, 183
    successive, 19–20
    visibility, 16–17
consequences of behaviors, 51, *see also* Behaviors
consumers, *see* Customers
contact points, friction, 99
contrived leaks, 162
control system, culture, 39
corporate strategy, 62
costliness for rivals, 15, 47–49
cost-plus pricing, 176
courage, 81–92, 108, 224
crisis, *see* Internal company conditions
cross-functional teams, *see* Teams
culture, *see also* Organizational culture and conditions
    anemic, 68–69
    concepts, 38–39
    innuendos, 23
    organization preparation, 37
    psychological dimension, 29–32
    strong *vs.* weak, 37
    thinking alignment, 172
customers
    concentration strategy implementation, 144–145
    creating, 211
    defining, 57–58
    internal company conditions, 127
    loyalty, 103
    problem-solving for, 58, 74, 195
    relationships and satisfaction, 35
cyber attacks, 189

D

databases, knowledge of employees, 45
dealer motivation, 180
deception, 188, *see also* Stealth activities
decisions
    Arthur D. Little Matrix, 136–138
    BCG growth-share matrix, 132–134
    ending point of campaign, 106–107
    General Electric Business Screen, 134–136
    management by objectives, 138–139
    overview, 131
    risk sensitivity, 15, 86
    Six Sigma, 139–140
decisive points
    concentration, 71
    correct identification, 142
    expansion campaigns, 12
    identification, 128
declining sales, reversing, 14–15
defender's resistance, weakening, 13–14
defenses
    building, 74
    competitive campaign, 216
    retaining share of market, 8–9
    vs. offense, 109–111
Dell, 96, 223
Delta airline, 197
demographic segmentation, 145
determination, leadership, 85–86
Dickens, Charles, 86
difficult markets, 96
direction, strategic plan, 55–60
direct strategy, *see* Indirect *vs.* direct strategy
disciplined judgment, 15
Disney, 95
dispersal, *see* Concentration *vs.* dispersal strategy
disrespect, *see* Respect
disruptive changes, 66–67, 172, 184
distractions, 14
distribution method changes, 149–150
distributors, 147–149
diversion, offensive move, 112
dogma, questioning, 172
dogs, 133
Dorf, Michael, 191

double agents, 161–162
doubts, *see* Fears
Drucker, Peter
　action, 115, 172, 213
　creating customers, 211
　management by objectives, 138
　zones of activity, 55
"drug delivery devices" example, 59–60
duration, competitive campaigns, 106–107, *see also* Ending point
Durcan, Mark, 91, *see also* Micron Technology

E

Eastman Kodak, 4, 53, 223
Eisenhower, Dwight D., 153
element of chance, 25–26
elements of strategies
　bold action *vs.* cautious restraint, 119–140
　concentration *vs.* dispersal strategy, 141–165
　indirect *vs.* direct strategy, 167–185
　overview, *xvi–xvii*
　valuing surprise and speed, 187–206
emotions
　baggage, 83–84
　deep-rooted, 98
　strength of mind, 89–90
employees
　aggressive competitor, 35
　communications, 36
　evaluating, 24–25
　fatigue, 24, 87, 100
　knowledge and skill databases, 45
　leadership relationship, 44–52
　market position, 35
　mindsets, 33–34, 54
　motivation, 7, 20
　offensive, 33–34
　overview, 32–33
　quality, 45
　reenergizing demoralized, 14–15
　self-development, 36
　technology, 35
　training, 20–22
　updating skills, 36
encircled markets, 96–97

ending point
　competitive campaign, 216
　determination, 112–114
　primary strategies, 72
　speed, 200–204
Enron, 40
Ericsson, 20
evaluation by exception, 149
existing market position, campaigns, 13
existing product promotion, 181
expansion into additional markets, 10, 12
Expectancy Theory, 43
expectations for plan, relationships
　active participation, 44–45
　competitive conditions, 50
　innovative thinking, 49
　maintaining momentum, 45–46
　market conditions, 50
　negative behavior, response, 51–52
　neutralization, competitor's strategies, 47–49
expendable agents, 162
expertise, area of, 56
explicit knowledge, 84
extrapolation, indirect *vs.* direct strategy, 172

F

Facebook
　central markets, 95
　Google's internal conditions, 105
　regulatory issues and industry trends, 151
　strength of character, 92
　successive confrontations, 19
failure
　application, 70
　as temporary setback, 72
　training and culture impact, 46
faintheartedness, 119
false moves, *see also* Stealth activities
　competitor evaluation, 23
　pretense to create fear, 46
　weakening defender's resistance, 14
false sense of security, 53
farm equipment example, 103
far-reaching focus, 171–172
fast-follower problem, 19

fatigue of employees, 24, 87, 100
favorable factor, 136
fears
    complacency, 66
    giving credence to, 101
    leader hesitation, 108
    offensive action importance, 71
    overview, 99–104
    postcampaign strategies, 65
    pretense to create, 46
    reduction, 72–75
    weakening defender's resistance, 14
Fiat Chrysler, 151
fight-or-flee mode, 18, 182
financing factor, 149
Five Guys, 158
five year plan, 56–57, *see also* Vision for future
flexibility
    chain of command reduction, 29
    disruptive change, 66–67
    pricing strategy, 177
    vs. inconsistent behavior, 90
Flipkart, 188
focus, far-reaching, 171–172
"fog of war," 65
follow pricing, 176
follow-up, campaigns, 112–114
Ford Motor company, 172
Frederick the Great, 184
friction
    competitive campaign, 215
    contact points, 99
    measuring, 100
    strategic plan and development, 72–75
    surprise element, 189

G

Galaxy Gear smartwatch, 20
Gates, Bill, 120
general agents, 160
General Electric
    company size, 223
    morale, 39–40
    organizational culture and employee input, 32–33
    Six Sigma, 139–140
    wearing-down process, 47

General Electric Business Screen, 134–136
General Motors, 27, 54, 168
geographic segmentation, 145, 146
Gerstner, Louis Jr., 74, *see also* IBM
global view, 195, 205
Goal Theory, 43
Google
    boldness, 121, 123
    competitive campaigns, 7
    conflicts, not isolated events, 19
    innovative thinking, 49
    internal conditions, 105
    offense, 111–112
    point of friction, 100
    regulatory issues and industry trends, 151
    tactical maneuvering, 173
Google Chromecast, 91
grueling work *vs.* whining/grumbling, 51
grumbling *vs.* grueling work, 51

H

Half Price Books, 184–185
Hart, B.H. Liddell, 168
healthcare example, 59–61, 63–65, 137–138
heart, 40–41
Heinz, 24
Henderson, Bruce, 133
Herzberg, Frederick, 42, 45
Hewlett-Packard, 223
Hierarchy of Needs, 42
Hollywood celebrities, 17
honor, 87–88
hospitality suite, 160
hostilities, 45
HTC, 20
human nature, 24
hypodermic needles, 59–60, 138, *see also* Healthcare example

I

IBM
    company size, 223
    dispersal strategy, 157
    ending point, 202–203
    friction reduction, 73–74

joint ventures, 150, 171
    strategic thinking, 169
Iger, Bob, 92, see also Disney
"Imagination Breakthrough" proposals, 33
Immelt, Jeffrey, 32–33, see also General Electric
implementation vs. planning, 102
increasing costs for rivals, 15
indirect strategies
    communication, 172–173
    cost-plus pricing, 176
    culture, thinking alignment, 172
    dealer motivation, 180
    dogma, questioning, 172
    existing product promotion, 181
    far-reaching focus, 171–172
    fear, 183–184
    flexible pricing, 177
    follow pricing, 176
    interpretation and extrapolation, 172
    loss-leader pricing, 177
    marketing, 178
    new product introduction, 180–181
    overview, xvi–xvii, 70, 167–169
    penetration pricing, 175
    phase-out pricing, 177
    preemptive pricing, 177
    pretest, plan evaluation, 220–221
    pricing, 174–177
    psychological pricing, 175
    question checklists, 221
    sale promotion, 180–181
    segment pricing, 176
    skim pricing, 175
    slide-down pricing, 176
    strategic thinking, 169–173
    stress, 182–183
    tactical advertising, 178–179
    tactic maneuvers, 173–181
    taking action, 172
    unbalancing competitors, 181–185
industry attractiveness, 134–135
industry trends, 151–152
inflexibility, see Flexibility
information factor, 149
in-house company newsletters, 161
innovative thinking, 49
inside agents, 160–161

Intel Corporation, 4–5, 53–54, 223
intellectual standards and performance, 97–98
intelligence, see Competitor intelligence
Intercloud, 169
intermediaries, 104, 146–150
internal company conditions
    competitor intelligence, 126–127
    full knowledge of, 23, 27
    Intel example, 54
interpretation, indirect vs. direct strategy, 172
In the Pink, 195
Intuit, 99–100, 102–103
intuition, 82, 83–85
iPad and iPhone, see Apple
isolated events, 5–6, 19–20, 224

J

J.C. Penney
    company size, 223
    defense vs. offense, 109–110
    friction, 101
    internal communications, 36
    organizational culture, 31–32
    tactical maneuvering, 173–174
Jobs, Stephen, 120
John Deere, 103
Johnson, Ron, 31–32, see also J.C. Penney
Johnson, Samuel, 86
Johnson & Johnson, 171
joint ventures, 10, 150
Joly, Herbert, 9, see also Best Buy

K

Kellogg, 24
Kenmore, 47
key markets, 94
knowing vs. reasoning, 83
knowledge management, 83–84
Kodak, see Eastman Kodak
Kotler, Philip, 194
Kun-hee, Lee, 19–20, see also Samsung Electronics

L

leadership
  absence, 203
  autocratic style, 40
  campaigns of opportunity, 10, 11
  central markets, 95
  challenging markets, 95–96
  character, strength of, 90–92
  competent, 80
  courage, 81–92, 119
  determination, 85–86
  difficult markets, 96
  encircled markets, 96–97
  faintheartedness, 119
  helping subordinates grow, 49
  honor, 87–88
  ineffectual, 69
  intellectual standards, 97–98
  intuition, 83–85
  key markets, 94
  leading edge markets, 93–94
  linked markets, 94–95
  market research, 152–153
  market selection application, 92–97
  mind, strength of, 89–90
  natural markets, 93
  organizational issues, 153–154
  overview, *xvi*, 79–83
  participative style, 40
  performance, 98
  planning, 153–154
  preemptive campaigns, 10
  presence of mind, 87
  pretest, plan evaluation, 214–215
  question checklists, 215
  reclusive and remote, 39
  reputation and recognition, 87–88
  roles, 22
  speed, 198–199
  staff relationships, 44–52
  training relationship, 22
leading edge markets, 93–94
leading factor, 137
leaks, contrived, 162
Lee, 142
LEGO, 15
lethargy, 67, 205
Levi Strauss & Co., 142–144, 210, 223

life cycles, 19, 193
limited-term *vs.* long-term campaigns, 14
linked markets, 94–95
Little (Arthur D.) Matrix, 136–138
live-and-let-live policy, 20, 46, 93
living agents, 162–163
Locke, Edwin, 43
Lollipop, 173
long-term objectives, 13
long-term *vs.* limited-term campaigns, 14
loss-leader pricing, 177
luck, *see* Chance
Lyft, 183

M

M&A, *see* Merger and acquisition route
maintaining momentum, 45–46
malaise, *see* Lethargy
management by objectives (MBO), 138–139
maneuvering, 194
Mann, Horace, 86
marketing, indirect *vs.* direct strategy, 178
markets
  central markets, 95
  challenging markets, 95–96
  concentration strategy
    implementation, 147
  conditions, 50
  difficult markets, 96
  encircled markets, 96–97
  expansion, 10, 12
  key markets, 94
  leading edge markets, 93–94
  linked markets, 94–95
  long-term in target, 74
  natural markets, 93
  new, long-term objectives, 13
  overview, 92
  position, employee input, 35
  realities, 155
  reclaiming former, 8
  research, leadership, 152–153
  retaining share, key region, 8–9
  segmentation bases, 145
  selection, 92–97, 147, 148
  share factors, 193
Maslow, Abraham, 41, 42, 45

Maslow's Hierarchy of Needs, 41, 42
material chain drawbacks, 28–29
matrices
    Arthur D. Little, 136–138
    BCG growth-share, 132–134
Mayer, Marissa, 16–17, *see also Yahoo!*
MBO, *see* Management by objectives (MBO)
MBR (management by results), *see* Management by objectives
McDonald's, 158–159, 223
McGregor, Douglas, 42–43, 45
McMillon, Doug, 80, *see also* Walmart
measuring friction, 100
mercury glass hospital thermometers, 138
merger and acquisition route, 95
Messenger, 92
Micron Technology, 95, 128
Microsoft
    dispersal strategy, 157
    Google competition, 7
    Google's internal conditions, 105
    regulatory issues and industry trends, 151
    Samsung competition, 19
    strategic thinking, 169
midlevel strategy, 62
mind, strength of, 89–90
Mini Cooper, 197
misinformation, 162
mistrust, *see* Trust
momentum, 45–46
Monsanto, 7, 30
morale
    business strategy impact, 41
    creating advantage, 105–106
    guidelines for maintaining, 34
    internal communications, 36
    leadership impact, 87
    organization preparation, 37, 39–41
    postcampaign strategies, 69–70
motivational theories, 37, 39–44
Motivation-Hygiene Theory, 42
motivation of personnel, 7
Motorola, 20, 139

N

Nadella, Satya, 91, *see also* Microsoft
natural markets, 93
negative behavior, response, 51–52, *see also* Behaviors
negotiation factor, 149
nerve, losing, 41
neutralizing competitors
    Best Buy example, 9
    boldness, 124
    competitive campaign, 216
    competitor's strategies, 47–49
    overview, 104–105
    physical and psychological characteristics, 17–19, 54
    primary strategies, 71–72
    success, 108
new markets, long-term objectives, 13
new product introduction
    concentration strategy implementation, 146–147
    sale promotion, 180–181
    tactical maneuvering, 173–174
new releases, 161
Nexus smartphone, 173
Nokia, 20, 157
noninterruption, 20–22
nonquantitative objectives, 60–61
nonviable factor, 136

O

Obama, Barack, 17
objectives
    competitive campaigns, 15
    strategies statement, 63–65
    strategy plan preparation, 60–61
    upward pressure, 15
obligatory commitments, joint-venture agreements, 10
Occupational Safety and Health Administration (OSHA), 151
offense and offensive
    characteristics, 111–112
    competitive campaign, 216
    employee input, 33–34
    primary strategies, 71
    vs. defense, 109–111
opportunities, 10, 11, 67, *see also* SWOT analysis
opposition, deliberately created, 45

ordering factor, 149
organizational culture and conditions, *see also* Culture
   boldness, 120–122
   leadership, 153–154
   pretest, plan evaluation, 210–211
   psychological dimension, 29–32
   speed, 199–200
organizational preparation
   active participation, 44–45
   competitive conditions, 50
   employee input, 32–36
   Expectancy Theory, 43
   Goal Theory, 43
   innovative thinking, 49
   maintaining momentum, 45–46
   market conditions, 50
   Maslow's Hierarchy of Needs, 42
   morale, 37, 39–41
   motivational theories, 37, 39–44
   Motivation-Hygiene Theory, 42
   negative behavior, response, 51–52
   neutralization, competitor's strategies, 47–49
   organizational culture, 29–32
   overview, *xv*, 27–28
   physical dimension, 28–29
   psychological dimension, 29–36
   relationships, expectations for plan, 44–52
   strong *vs.* weak cultures, 37
   Theory X/Theory Y, 42–43
   Theory Z, 43
organizational structure, culture, 39
Ouchi, William, 43, 45
overestimation of strength, 24

P

Page, Larry, 92, 120, 123, *see also* Google
Paley, Norton, 227–229
Palmisano, Sam, 74, *see also* IBM
Panasonic
   available reserves, 60
   company size, 223
   turnaround conditions, 3–4, 53
Paper, 92
paradigm, culture, 39
participation, active, 44–45
participative leadership, 40
passive resistance, 48–49
patterns, competitive operating, 125
payment factor, 150
PayPal, 193
penetration pricing, 175
performance, leadership, 97–98
personnel, *see* Employees
phase-out pricing, 177
physical dimension
   building defenses, 74
   expected confrontations, 16–17
   neutralizing the competitor, 17–19
   noninterruption, 20–22
   organization preparation, 28–29
   successive confrontations, 19–20
physical energy, *see* Fatigue of employees
physical possession factor, 149
planning
   lack of skills, 191
   leadership, 153–154
   vs. implementation, 102
positioning
   branding, 197–198
   flexible work teams, 194
   global view, 195
   long-term, 74
   maneuvering, 194
   overview, 194
   problem-solving for customers, 195
   protection, 16
   solidifying, 13
   strategy development, 195–196
possession, 14–15, 149
postcampaign strategies
   application failure, 70
   competitor intelligence, inadequate, 68
   complacency, 66
   culture, anemic, 68–69
   inflexibility, disruptive change, 66–67
   leadership, ineffectual, 69
   lethargy, 67
   morale, 69–70
   overview, 55, 65–66
   resource dispersal, 67
Potts, Randy, 121–127, 128
Powell, Colin, 86
power structure, culture, 39
preemptive actions, 10, 177

presence of mind, 87
press releases, 161
pretest, plan evaluation
    bold action, 217–218
    competitive campaign, 215–217
    concentration *vs.* dispersal strategy, 219–220
    conditions triggering turnaround, 210
    indirect *vs.* direct strategy, 220–221
    leadership, 214–215
    organizational condition, 211
    overview, *xvii,* 209
    strategy plan, 212–213
    surprise and speed, 221–223
pricing
    avoiding price wars, 177
    indirect *vs.* direct strategy, 174–177
    lethargy, 67
primary strategies
    concentration, decisive point, 71
    ending point, campaign, 72
    indirect strategies, 70
    neutralization, competitor, 71–72
    offensive, 71
    resources, application, 71
    speed, 70–71
problem-solving for customers, 58, 74, 195, *see also* Customers
process culture, 38
procrastination, 70–71, *see also* Speed
product attributes segmentation, 145
product introduction, 146–147, 180–181
product literature/specification sheets, 161
product promotion, existing, 181
"product systems" example, 59–60
psychographic segmentation, 145
psychological dimension, *see also* Behaviors
    aggressive competitor, 35
    Best Buy example, 9
    boldness, 34, 218
    building defenses, 74
    communications, 36
    difficulty interpreting, 105
    employee input, 32–36
    expected confrontations, 16–17
    indirect strategy, 220
    market position, 35
    neutralizing competitors, 17–19
    noninterruption, 20–22
    offensive, 33–34
    organizational culture, 29–32
    overview, 29, 32–33
    pricing, 175
    successive confrontations, 19–20
    surprise element, 190
    technology, 35
    weakening defender's resistance, 14

Q

Qualcomm, 151
quantitative objectives, 60–61
question checklists
    bold action, 218
    competitive campaign, 217
    competitive crisis *vs.* random occurrence, 5–6
    concentration *vs.* dispersal strategy, 219–220
    endpoint of campaign, 113–114
    indirect *vs.* direct strategy, 221
    leadership, 215
    organizational conditions, 211
    strategy plan, 213
    surprise and speed, 222–223
    turnaround plan, key areas, 210
questioning dogma, 172
question marks, 133

R

random occurrence *vs.* competitive crisis, 5–6, 19–20, 224
readiness, 21
realities, marketplace, 155
reclaiming former market, 8
recognition, 87–88, *see also* Rewards and recognition
recreational vehicles example, 121–122
Redmi 1S smartphone, 188
reenergizing demoralized personnel, 14–15
regulatory issues, 151–152
Reis, Al, 194
relationships
    active participation, 44–45
    competitive conditions, 50

238 • Index

innovative thinking, 49
maintaining momentum, 45–46
market conditions, 50
negative behavior, response, 51–52
neutralization, competitor's strategies, 47–49
solidifying, 146
reputation, 87–88
reserves, use of, 114–115, 216
resistance, 13–14, 72
resources, *see also* Concentration *vs.* dispersal strategy
application, 71
big picture importance, 94
lack of, preemptive campaigns, 10
unnecessary dispersal, 67
respect, lack of, 69–70
restraint, *see* Bold action *vs.* cautious restraint
retaining share of market, key region, 8–9
reverse declining sales campaigns, 14–15
rewards and recognition, 45, 69, 87–88
Ricoh, 8, 110
risk, 15, 86, 96
risk taking factor, 149
rituals, culture, 31, 38, 39
Roku, 91
Rometty, Virginia, 74, 151, 154, 202, *see also* IBM
Roosevelt, Franklin D., 86
root causes, trigger identification
chance, element of, 25–26
competitive campaign types, 7–15
expansion into additional markets, 10, 12
expected confrontations, 16–17
increasing costs for rivals, 15
limited-term *vs.* long-term, 14
neutralizing the competitor, 17–19
new markets, long-term objectives, 13
noninterruption, 20–22
obligatory commitments, joint-venture agreements, 10
opportunity, 10
overview, *xv*, 223
physical/psychological dimensions, 16–22
preemptive, 10
primary activation conditions, 3–7

reclaiming former market, 8
reenergizing demoralized personnel, 14–15
retaining share of market, key region, 8–9
reversing declining sales, 14–15
solidify position, 13
standstill factors, 22–26
successive confrontations, 19–20
upward pressure, ambitious objectives, 15
weakening defender's resistance, 13–14
routines, culture, 39
Royal Dutch Shell, 200
rumors, 161, *see also* Customers; Employees
ruses, *see* False moves

S

sale promotion, 180–181
sales, reversing declining, 14–15
Salk, Jonas, 83
Samsung Electronics
caution, 128
conflicts, not isolated events, 19–20
dispersal strategy, 158
ongoing competitive activity, 20
Panasonic comparison, 4
speed, 188
surprise element, 190
wearing-down process, 47
Schultz, Howard, 24, *see also* Starbucks
secrecy, 188
security, 188–189
segment-by-segment approach, 156
segment pricing, 176
self-centeredness, 91
self-control, 89, 120
self-sabotage, 79–80
sensitivity, 15, 86
Sharp, 110
short life cycles, 19
Simon, Bill, 80, *see also* Walmart
SIMS, *see* Smart Inventory Management System
Six Sigma, 139–140
SK Hynix, 128
skim pricing, 175

slide-down pricing, 176
sluggishness, *see* Lethargy
Smart Inventory Management System (SIMS), 172
smartphones and smartwatches, *see specific brand*
Smith, Brad, 100, 103
social distinction, 88
Soft-Layer Technologies, 151
SolarCity, 182
Southwest Airlines, 197–198
speed, *see also* Surprise and speed
    barriers to implementation, 198–204
    ending point, 200–204
    leadership, 198–199
    organizational conditions, 199–200
    primary strategies, 70–71
Spielberg, Mark, 120
Square, 193
staff, *see* Employees
standards, 97–98
standstill factors, 22–26
Staples, 223
Starbucks, 24
stars, 133
statement, strategic plan, 63–65
status, 88
stealth activities, 162, *see also* Agent utilization; Deception; False moves
stories and myths, culture, 31, 39
strategic plan and development, *see also* Turnaround plan and development
    comparative analyses, 72
    competitor intelligence, inadequate, 68
    competitor neutralization, 71–72
    complacency, 66
    concentration, decisive point, 71
    corporate strategy, 62
    culture, anemic, 68–69
    direction, 55–60
    ending point, campaign, 72
    failure, 70, 72
    friction reduction, 72–75
    importance, 225
    indirect strategies, 70
    inflexibility, disruptive change, 66–67
    leadership, ineffectual, 69
    lethargy, 67
    midlevel strategy, 62
    morale, 69–70
    objectives, 54, 60–61
    offensive, 71
    overview, *vv*, 53–55
    positioning, 195–196
    postcampaign strategies, 55, 65–75
    pretest, plan evaluation, 212–213
    primary strategies, 70–72
    question checklists, 213
    resistance, circumventing, 72
    resources, 67, 71
    speed, 70–71
    statement, 63–65
    strategies, 62–65
    supporting strategies, 72–75
    tactics, 63
    vital factor, 225
strategic thinking, 169–173
strengths, *see also* SWOT analysis
    business, 135–136
    gain factors, 201
    loss factors, 202
    overestimation of, 24
strong factor, 137
strong *vs.* weak cultures, 37
structure, competitive campaigns, 99–115
stubbornness, 90–91
Subway, 158
success, *see also* Pretest for evaluation
    campaign follow-up, 112
    exploiting initial, 109
    measuring, 20
    speed, 70–71
successive confrontations, 19–20
suggestions, lack of procedure, 70
Sun Tzu
    breaking resistance without fighting, 184
    competitor intelligence, 21
    indirect *vs.* direct strategy, 167, 168, 220
    prolonged warfare, 205
supply chain and partners, 146, 189
supporting strategies
    circumventing resistance, 72
    comparative analyses, 72
    failure as temporary setback, 72

friction reduction, 72–75
surprise and speed
    barriers to implementation, 198–204
    branding, 197–198
    ending point, 200–204
    flexible work teams, 194
    global view, 195
    leadership, 198–199
    maneuvering, 194
    organizational conditions, 199–200
    overview, *xvii*, 187–191
    positioning, 194–198
    pretest, plan evaluation, 221–223
    problem-solving for customers, 195
    question checklists, 222–223
    speed, 191–194
    strategy development, 195–196
    valuing, 204–205
SWOT analysis, 154, 163–165
symbols, culture, 31, 39
syringes, 138, *see also* Hypodermic needles

T

tacit knowledge, 84
tactical advertising, 178–179
tactic maneuvers, 173–181
tactics, strategic plan, 63
teams
    cross-functional, 32, 67
    duties and responsibilities, 30
    flexibility, positioning, 194
    ineffectual leadership, 69
    lacking team spirit, 70
technical papers, 160
technology
    customer satisfaction, 58
    employee input, 35
    friction reduction, 73
tenable factor, 136
termination, *see* Ending point
Theory X/Theory Y, 42–43
Theory Z, 43
thinking, 49, 169–173
threats, *see* SWOT analysis
Time Warner Cable, 35
title factor, 150
togetherness, uncharacteristic, 93

tools, *see* Decisions
Toshiba, 128
tough-guy macho culture, 38
Toyota, 168
trade shows, 160
training, 20–22, 69
transient events, 5–6, 19–20, 224
transitions, distributor changes, 147–149
Trout, Jack, 194
trust, 22, 69, 84
turnaround plan and development,
    see also Strategic plan and development
    chance, element of, 25–26
    competitive campaign types, 7–15
    expansion into additional markets, 10, 12
    expected confrontations, 16–17
    increasing costs for rivals, 15
    limited-term *vs.* long-term, 14
    neutralizing the competitor, 17–19
    new markets, long-term objectives, 13
    noninterruption, 20–22
    obligatory commitments, joint-venture agreements, 10
    opportunity, 10
    organization preparation, 27–52
    overview, *xv*
    physical/psychological characteristics, competitive conflicts, 16–22
    preemptive, 10
    primary activation conditions, 3–7
    purpose, 7
    question checklists, 210
    reclaiming former market, 8
    reenergizing demoralized personnel, 14–15
    retaining share of market, key region, 8–9
    reversing declining sales, 14–15
    solidify position, 13
    standstill factors, 22–26
    strategy plan preparation, 53–75
    successive confrontations, 19–20
    upward pressure, ambitious objectives, 15
    weakening defender's resistance, 13–14
Tzu, *see* Sun Tzu

U

Uber, 168–169, 183
Ullman, Mike, 32, *see also* J.C. Penney
unbalancing competitors
    campaign components, 104
    indirect strategies, *vs.* direct strategy, 181–185
    offensive moves, 112
    psychological advantage, 74
unexpected conflicts, 16–17
unintended possibilities/probabilites, 25–26, *see also* Chance
United airline, 197
universals, 70
unseen elements, culture, 38
unserved market niches, 184
upward pressure, ambitious objectives, 15, *see also* Risk
utilization
    agents, 159–163
    concentration strategy, 155–156

V

valuing surprise and speed, *see* Surprise and speed
vigilance, *see* Awareness
vision for future, 61, *see also* Five year plan
Volkswagen, 151
von Clausewitz, Carl, 168
Vroom, Victor, 43
vulnerabilities, 141, *see also* Concentration *vs.* dispersal strategy

W

wait-and-see approach, 18
Walmart, 7, 79–80
Watson system, 151, 171, 203
weakening defender's resistance, 13–14
weak factor, 136
weaknesses, *see* SWOT analysis
wearing-down process, 47
Welch, Jack, 39–40, 120, 139, *see also* General Electric
WhatsApp, 92
whining *vs.* grueling work, 51
Whirlpool, 47
Winfrey, Oprah, 17
Winnebago Industries, 121–127, 128–129
winning spirit, 41
work-hard, play-hard culture, 38
work teams, *see* Teams
Wrangler, 142
Wright, Sharon Anderson, 185

X

Xerox
    defense *vs.* offense, 110–111
    expansion campaigns, 12
    leading edge market, 94
    limited-term *vs.* long-term campaigns, 14
    reclaiming former market position, 8
    solidifying existing market position, 13
Xiaomi, 20, 188

Y

Yahoo!, 16–17
YouTube, 92

Z

zones of activity, 55
Zuckerberg, Mark, 92, *see also* Facebook

About the Author

Norton Paley has brought his world-class experience and unique approach to business strategy to some of the global community's most respected organizations.

Having launched his career with publishers McGraw-Hill and John Wiley & Sons, Paley founded Alexander-Norton Inc., bringing successful business techniques to clients around the globe, including the international training organization Strategic Management Group, where he served as senior consultant.

Throughout his career, Paley has trained business managers and their staffs in the areas of planning and strategy development, raising the bar for achievement and forging new approaches to problem solving and competitive edge.

His clients include:

- American Express
- IBM
- Detroit Edison
- Chrysler (Parts Division)
- McDonnell-Douglas
- Dow Chemical (Worldwide)
- W.R. Grace
- Cargill (Worldwide)
- Chevron Chemical
- Ralston-Purina
- Johnson & Johnson
- USG
- Celanese
- Hoechst
- Mississippi Power
- Numerous midsized and small firms

Paley has lectured in The Republic of China and Mexico and he has presented training seminars throughout the Pacific Rim and Europe for Dow Chemical and Cargill.

About the Author

As a seminar leader at the American Management Association, he conducted competitive strategy, marketing management, and strategic planning programs for over 20 years.

Published books include:

- *The Marketing Strategy Desktop Guide*, 2nd edition (Thorogood, 2008)
- *How to Develop a Strategic Marketing Plan* (CRC Press, 1999)
- *The Manager's Guide to Competitive Marketing Strategies*, 3rd edition (Thorogood, 2005)
- *Marketing for the Nonmarketing Executive: An Integrated Management Resource Guide for the 21st Century* (CRC Press, 2000)
- *Successful Business Planning: Energizing Your Company's Potential* (Thorogood, 2004)
- *Manage to Win* (CRC Press, 2004)
- *Mastering the Rules of Competitive Strategy: A Resource Guide for Managers* (Auerbach Publications, 2007)
- *Big Ideas for Small Businesses* (Viva Books, 2011)
- *How to Outthink, Outmaneuver, and Outperform Your Competitors: Lessons from the Masters of Strategy* (Productivity Press, 2013)
- *Clausewitz Talks Business: An Executive's Guide to Thinking Like a Strategist* (CRC Press, 2014)

On the cusp of the interactive movement, Paley developed three computer-based, interactive training systems: The Marketing Learning Systems; Segmentation, Targeting, & Positioning; and The Marketing Planning System.

Paley's books have been translated into Chinese, Russian, Portuguese, and Turkish.

His byline columns have appeared in *The Management Review* and *Sales & Marketing Management* magazines.

Excerpts from reviews of books by Norton Paley:

A book too forceful to ignore.
> Review in *Business Line*, financial daily for The Hindu Group of Publications about Paley's *Manage to Win*.

This book is both intellectual and practical ... an interesting vehicle for presenting detailed planning concepts ... it is clear and well-organized.
> T. J. Belich, in *Choice*, about Paley's How to Develop a Strategic Marketing Plan.